To Paul.
From Julianna
Christmas '80

THE CIRCUIT

The

Ralph M. Demers

Circuit

A RICHARD SEAVER BOOK

THE VIKING PRESS   NEW YORK

Copyright © 1976 by Ralph M. Demers
A Richard Seaver Book/The Viking Press
First published in 1976 by The Viking Press, Inc.
625 Madison Avenue, New York, N.Y. 10022
Published simultaneously in Canada by
The Macmillan Company of Canada Limited
Printed in U.S.A.

Library of Congress Cataloging in Publication Data
Demers, Ralph M
   The circuit.
   "A Richard Seaver book."
   I. Title.
PZ4.D3766Ci [PS3554.E457]   813'.5'4   76-11833
ISBN 0-670-22268-2

Thanks to the village of Jacmel, Haiti,
and the Tennis Club Vallarta, Puerto Vallarta, Mexico

To Valerie Jenkins

THE CIRCUIT

1

The almost inaudible hum of the 747 sent a smooth vibration through the darkened compartments. In the dim emergency-light glow, the two forms were locked together in silhouette. A sleepy passenger fumbled by in a half-trance and Shep took a step backward in the galley, gave the girl's hand a parting squeeze, and returned to his seat. He stepped over the racquets on the floor; six of them, bound together with masking tape to make them more manageable. With first a finger, then his whole hand, he caressed them, then leaned back and closed his burning, sleepless eyes.

The rain was holding back until nightfall, but Heathrow Airport was stickily misting over as they put down.

"Not a wet Wimbledon, I pray God. Not this one," he thought, as he ducked through the door to the ramp. *His* hostess was showing red, Kewpie-doll lips and freshly brushed teeth as she said her sweet good-byes and "pleasure-to-fly-with-yous," and to Shep an extra flutter of her lashes with "See you again very soon" in her eyes. His lying nod of acknowledgment moved him past her and into the terminal. Quickies didn't make for good long-term relationships.

He filled out the passport control form:

NAME—Francis Shepard

AGE—30

PLACE OF BIRTH—Laguna Beach, California, USA

For SEX, he used to fill in "Occasionally," and for FIRM, "Yes," but it always entailed making a new form out and he didn't have the time or inclination for childishness this trip. The official made the usual inane comments about the racquets.

The other passengers who had been on the same overnight flight from New York felt possessive, protective of him as though he were their representative and contribution to the sports world. Bags were tumbling down the conveyor belt onto three roundabouts, and he recognized some of his fellow passengers gathering around the center one. Then, one of the mini-miracles that makes Wimbledon the greatest tournament of them all manifested itself, one of the thousand little sweeteners so startling to first-timers and very much looked forward to by the "boys." A chauffeur-capped man in black already had Shep's tennis bag and Vuitton case in a supermarket carriage and was beckoning him toward the customs inspection line. Of course, not all players got that kind of treatment, but Shep, after years of drifting along as a world-class rabbit, never winning the big ones, fodder for the edges of the "long knives" on their path to the later stages, finally was a winner. He would have been very surprised indeed not to be third or fourth seed when they were announced during the Queen's Club London Grass Court Tournament, played immediately before Wimbledon.

Professional tennis players no longer dressed like hippies or gypsies as they did while "amateurs." Shep wore a beige silk turtleneck sweater, a brown suede safari jacket, and hip-hugging tan corduroy bell-bottom trousers, all of which could have made him look like a screaming faggot except for the strong, straight, athletic movement of his body. He was tall and walked tall.

"Thanks very much for grabbing my bags. How did you know which one I was?" he asked.

"Oh, I've seen you quite often on the telly, and your face is all over the sports pages here these days, sir," the livery man answered in heavily accented Londonese—not quite Cockney, not quite plummy.

As they sped along the M4 toward the city, Shep leaned forward in the Rolls and asked through the chauffeur's window, "Is this rain going to keep up all month, or what?"

"Oh, no, sir. We have a saying. If you don't like the weather in London, just wait a minute." Five minutes later, as they roared along the Chiswick flyover, above the long rows of grey houses, council flats with their well-worked and well-loved little gardens, the sun exploded through the billowing grey damp, and the whole tableau of London was limned in brilliant color.

His reverie was interrupted by the driver's repeated questions.

"That was the Gloucester Hotel, wasn't it, sir?"

"What? Oh, yes, please."

"Will you be going back out to the courts straight away?"

"No, I'm going to shower and change first. I won't be needing you anyway, though. I know my way around pretty well. I'll probably go out later by tube."

They motored down Cromwell Road past Baron's Court, where the Queen's Club was, and Shep started to get that tight, hollow, gnawing feeling in his gut. Whether fear or heightened anticipation, it always struck him before a major tournament. He'd be practicing there later that day, and hitting the ball would release him from the torment temporarily. Until the big moment.

More than the normal hectic movements of a busy hotel—the ebb and flow of check-ins and -outs, tour groups, hookers going home with their little bags—the flavor of the Gloucester lobby was show-business busy. Tennis players were cannoning from elevator doors, surging in and out and around the dining-room floor and lounging everywhere.

Shep passed up breakfast in the hotel in order to get a quick start on the beautiful morning. His original plan had been to stroll a bit, collect himself, and do some shopping. Get the feel of London again. He was usually quite at peace here. The tube station was across the street, though, so he made his way among the smiling throngs, the flower carts and news vendors, and entered the station, his thoughts myriad—disorganized, but light. Queen's Club was only two stops away but they were long ones. Shep wasn't completely awake and his thoughts drifted into the alpha areas, warm and quasi-conscious. At one point he saw his co-passengers as functioning, transparent bodies with their circulation and respiratory systems all in motion. When their bowel systems began to get into gear he snapped out of it, and the train rolled up to his station.

The four blocks from Baron's Court Station to the club were very plus-charged. The delicious morning air combined with the heady wine of challenge to produce an exhilarating tonic. He strode happily through the gates, racquets under one arm, bag swinging in a high pendulum from the other. Shep was always startled by the sudden transition from city to verdant spaciousness. This massive green oasis right in the heart of London was so incongruous it always upset his sense of balance at first impact.

Queen's Club was a living anachronism. The largest improvement to the enormous tennis complex had taken place forty years before, when the Nazis had inadvertently bombed indoor courts four and five. They had been aiming for Montgomery's headquarters, billeted in St. Paul's school in a similar grassy environ a quarter mile away. Since then, the club had muddled through with paintless, spartan dressing rooms, an unimaginative bar that looked forever like a second-class waiting room in a Turkish railway station, and four toilets, somewhat inadequate during tournament time—one hundred twenty-eight competitors, and some of them mighty uncomfortable.

It had long ceased to belong to the membership. The LTA had bought it at a rock-bottom bargain, keeping it from sinking altogether into the Red Sea of ink, but firmly grasping it in its cheapthinking, self-serving talons. Fifteen grass courts, twelve of them tournament quality, twelve composition (hard courts, the English call them), three synthetic, five indoor wood, five squash-racquet, one racquet, and two real tennis courts—all serviced by five grounds keepers who managed by herculean energies, genius, and good humor to maintain the entire complex. Very nearly, anyway.

A large weeping willow (weeping for former splendor) stood amid the grass courts. All of the courts, separated by hip-high canvas walls, were frantic with the feverish practice of the young Wimbledon dreamers. Shep, the great Australian Jesse Fraser, and some of the other veterans had booked the indoor wood courts in case of rain (and, of course, it did rain later). The wood was very fast, the fastest tennis surface in the world, and more resembled the fast grass of Wimbledon than the comparatively slower stuff at Queen's. The Queen's tourney, although important in its own right for prize money and points, was used essentially by the big guns as a warm-up for Wimbledon.

The tearoom was bustling—noisy, happy, electric with good vibes. Shep waved and made small talk with friends at tables, in the cafeteria line, and as they were tearing by with racquets. Some he'd seen only days before in a team tennis match, others during the World Championship Tennis Finals a month before; still others, who played other circuits, he hadn't been together with for almost a year. All highly competitive yet friends, some in word and some in depth and deed. Randy Mariano pumped the arm that was full of racquets, blustering hellos with phoney intensity. Randy, a fellow Californian, bullshitted him right in the face with, "Hey, baby— hey, Shep—man! It's great to see you. Where're you staying?" It seemed to be categorical that all big-time tennis players were attractive, and Randy, with masses of tight blond curls, was no exception. Only the slight twist of his hard-bitten, cynical smile marred the fair Piedmontese face.

Shep wriggled his arm free with the ploy of unloading his racquets onto an empty chair. "I'm at the Gloucester with the rest of the circus." The tone of Shep's answer wasn't intended to inspire further conversation. Randy was a prick, and Shep treated him like a parking meter without any more time on it. Shep moved to the end of the breakfast line and Randy hopped gingerly right in behind him, continuing, "Far out! Any pussybilities over there? I'm at the Westbury and it's quiet, man! Really."

Shep half shrugged his "I wouldn't know" punctuation to the conversation and immediately involved himself in a genuine reunion with Jesse Fraser and Missy Teaford ahead of him in line. Missy had been in and out of Shep's life for years, and their vibrations for each other were mixed and complex. She was smiling impassively at him now with her tanned cheerleader face, which glowed with skin moisteners that were at work on the smile lines around her eyes and mouth. The sun, the great destroyer.

He and Jesse were worlds apart in nationality and philosophies.

The English think Australians and Americans are similar. They aren't. But Jesse and Shep had other common denominators: Jesse, a former winner at Wimbledon, and Shep, just months before, the WCT champion. They were keen, unyielding opponents who had faced each other so many times in high-voltage, dramatic situations. Yet, without being confidants or engaging in a sham friendship pattern, they had a warm relationship, a comfortable, workable one. They were actually glad to see each other, even though it would probably come to a showdown between them somewhere in the later stages of the big one.

Al Wick was sitting with a group of fellow Australians at the end of a large table. His features were heavy, yet strangely patrician, and the broken tooth right in front gave him a boyish, rascally look. And Wicko knew it.

Trays were moved around as room for Jesse and Shep was made at the table. Shep became the international Mason-Dixon Line between the Aussies and a cluster of Americans. Randy pushed in beside him, making what could have looked like a pitiful attempt at acceptance—to anyone but Randy. When Missy saw Wicko at the table, she moved off quickly to another part of the room, after sweet little parting mumbles.

There was casual warmth in the group as they talked about tennis, past and present. Only Randy talked about himself. The conversation was, otherwise, careless, useless, and good-natured.

Consuelo Alvarro, the world number one from Argentina, floated gracefully past the table, chatting over her shoulder to the notorious Angie Redfield. Wick leaned over and whispered something in Shep's ear, and Shep replied with an "I don't think so" shake of his head. Wicko went back to an argument that was in halting progress, interrupted by comings and goings and the great waves of noise that rolled through the place. Freddie Moore, a rangy, red-headed Aussie, was amusing himself by getting annoyed on the subject of the number-one male player in the world, the probable first seed at Wimbledon, Franco Berconi. "I think Berconi is a proper nut case," Freddie was saying. "He goes berserk for the slightest reason and throws everybody's concentration off. He's not just a bad sport, you know. That's cheating, in a way."

"Berconi's a winner," Randy chipped in. "In my opinion, sportsmanship is for losers."

"Opinions are like assholes, Mariano," Wick said. "Everybody has one."

"Well, he's a winner, and that's a fact," Randy shot back. "Opinions based on fact are a lot more valid, man."

"Bullshit!" Wicko explained. "Facts are like bricks, mate. You can build things with them or just break windows." He was quite pleased with that one.

"You're out of your tree, man. Have you ever in your life beaten a fit guy? They all have excuses. Winners don't have to explain."

Jesse winced at Randy's last remark and stood up. He had beaten Randy in fourteen of their sixteen meetings. "Well, Wicko," he said, "let's suit up and get some practicing in or we'll all be on the next plane out of here." He moved toward the door with a slight shuffle, his narrow, sinewy shoulders slung in their characteristic boxer's droop. The group rattled chairs and grabbed bags, racquets, and notebooks—the last a new innovation for an orderly building up of confidence with a progress check-list.

As they passed by, Laurie Silverman looked up and pushed aside a swath of ebony-luster hair that gleamed with health and recent washing. Even amid the din, the excitement of young people swimming through the noisy vibrations of their own enthusiasm, Laurie was sheltered in the velvet of her very private dreams. She looked at them. When she didn't see Shep among them (he'd already gone upstairs), she lost interest and resumed jabbing at her fruit salad. If she were annoyed by the half a cherry that squirted out and stained her new Teddy Tingling tennis outfit, it couldn't be read on the peaceful planes of her face. Her composure was real enough. Laurie had no hang-ups about being Jewish in America, American in Europe, or a woman in a predominantly man's world. Her only hang-up was her dedicated and illogical, total and undeclared love for Francis Shepard. It was like religion with her: one believed or one didn't. She collected his press clippings and watched his interviews and matches on the box every chance she had. Even as her long, slender fingers dipped water from her glass and dabbed the little handkerchief at the spot on her dress, she thought of Shep. The intent expression smoothed into a smile. She knew it was a bit crazy . . . and so what!

In the only other backwater in the human, tear-ass, white rapids of a tearoom sat Mikhail Pakachev. His stolid unconcern for the wild bustle around him was less composure than the inability to communicate. Mikhail was a warm, emotional person. The passive sense of humor that played about his large black eyes was a give-away. He was a sounding board—amusement waiting to be triggered. At least when he could understand what the hell was being said. Unlike his predecessor Alexis Metreveli, Pak's English was terrible. So he sat there alone and lonely. Lonely for his cramped apartment, his dumpy but pretty wife, and his beautiful little son. In his hometown on the Black Sea he was a hero, all bounce and laughter. Here? Here he sat, staying small and quiet in this bedlam of friendly noise. After what had recently happened to him in Russia, he shunned the company of other Russians. Better to be alone and safe and miserable.

Upstairs, Wick and Jesse were changing. Wick was all brawn, Jesse a study in angles.

"Ya know, Jess, Berconi's not so bad. I kinda like him off the court. He's funny. A bloody riot, really. When he says 'Geta stuff-ed,' for instance, that sort of thing. Funny."

"I know, mate. He just thinks he's the reason he's so tall."

"Did you see Missy frost me? I don't understand her," Wick mused. "She's a real enema to me."

"You mean enigma, don't you? An enema's when they give you a shot in the ass with a hose."

"That's her," Wick laughed. "And what a dung-tongue. I can't stand a foul-mouthed bitch."

"But I bet they just adore you, you sweet-talking mother. Maybe she's heard some of the wonderful, generous things you say about her, eh?"

Jesse always took special care of his feet, and put his two pairs of socks on very carefully before pulling on the Adidas shoes. His body was his money-making machine and he took care of it. He finished lacing them and started off down the stairs. "Grab six new balls, will you, Wicko? I have to make a piss-stop. We're on court four," he shouted back up. "Freddie's already down there."

After an hour of two-on-one indoors in the summer heat they were dripping wet and beginning to slip in their own perspiration. One at the net volleying (at the moment, Jesse), the other two, with their free hands and pockets full of balls, firing shots at him machine-gun fashion. Trying passing shots, topspin lobs, cross-courts; and Jesse, leaping about, smashing, crunching backhand volleys—one of his deadly shots—drilling forehand volleys up the middle, running back to retrieve a lob with a backhand lob of his own, and they, hitting another ball out of their hands when his sailed out, then taking the net and volleying to both sides of the doubles court, running him all over with deep, crisp shots. No resting except for a moment when another man (this time, Wick) would become the single man and the process would continue.

Finally, Wick slipped a little on a wet spot on the boards and called a time out to get some sawdust from the umpire's chair.

"Chroist," he panted, "what's the expression? I feel as busy as a one-legged man at an ass-kicking contest! Hey, Jess, I'm having . . . I'm having a problem with my forehand volley—on the high ones especially. Can you see what I'm doing to fuck it up?"

"We still have five minutes," Jesse said, looking at the clock on the balcony. Jaguar Gray and the English number three, Billy Sherman, were up there talking. They were next on court.

"A high volley's a bit of a problem. It's one of the two toughest shots for me. The backhand smash is the other. The normal volley's a reflex action for me—a rhythm conditioning. But watch . . ." Jesse went through a pantomime of dragging his right foot and snapping his wrist. "Usually you've got to stand there waiting for the ball and your timing gets thrown off. So if it's high enough, I try to get a little stutter into my motion and slide under it and . . ." he went into the motion again ". . . smash it with lots of wrist. If I have to volley—if it's just that little bit too low to crack, I think 'out'—never down. Don't hit down at it. Little more backswing than usual, firm wrist and—out. It's the footwork, though, that's the thing. The little drag I do makes up the time. You have to move in on it, Wicko. You're waiting for it. It's tough. Serve some more and come in. I'll try to hit a couple of them."

Through all of this, Freddie Moore was lying down on the cool, green wooden floor. His upturned eyebrows formed his face into a perpetual question, even with his eyes closed.

"Freddie, get up and piss off out of there and let me have a go at this flaming volley," Wick said, nudging Moore with his foot.

Jesse waved to the two in the balcony, who were just then being joined by Shep and another player Jesse didn't recognize.

"We'll only be a couple more minutes, Jago," he called up. The group on the balcony nodded in acknowledgment and went on with their whispering.

After several trips to the net to sort out Wick's problem, Jesse and the rest put on their track suits and gathered up their gear.

"What a furnace it is in here," Shep exhaled loudly. He fanned himself with a covered racquet, saying, "The sky's getting very black out there. Just fat and heavy with rain . . . and soon."

The air in the court was close and musty. Shep, Jaguar, and Billy Sherman kept their warm-up pants on but removed their jackets. A slightly overweight American was introduced to the Australians as Sammy Kerwin, once a junior champion and now trying his now-or-never shot at the big time. Sammy wasn't playing Queen's because the Wimbledon Qualifications at Roehampton conflicted with it. Ten berths at Wimbledon were held open for the fortunate few who emerged from the grueling, heartbreaking competition. Kerwin kept his suit and sweater on in a heroic effort to drop the weight before Monday—two days away. At twenty-nine, it was going to be tough enough even fit, amidst the hot breath and bared fangs of the hundred young bloods competing at Roehampton.

"Let's drop our stuff off upstairs and go for a run," Jesse suggested. The grass on the fifteen courts had been freshly mown the night before and the crisp, green aroma was delicious and inspiriting. They jogged the necessary number of laps to make up two miles and sprinted the final two hundred yards.

Lightning brightened the soundless sky off to the northwest as they trotted up to the front of the clubhouse. Randy was standing on the steps with his racquets, apparently waiting for them.

"Hey, Wicko, man, I've got an indoor court booked now if you've got anything left. I'd like to hit some."

"The name is Mr. Wick to you, Mariano. And if I want to play with a prick, I'll play with my own."

They all laughed, including the seemingly insensate Randy, who said, "That's beautiful! You're out of sight, Mr. Wicko!" Randy's sense of humor was at once a buffer and his only saving grace.

Wick ignored him and followed the others up the steps just as the first heavy drops began to fall. The outside courts suddenly looked like a disturbed anthill as everyone ran here and there, picking up their things and scrambling for cover. Jesse stood quietly for a moment, luxuriating in the warm, soft-scented summer rain.

The hansom cab hung precariously from the ceiling of the pub, not too strangely called the Hansom Cab. Jaguar Gray was holding court among the groupies and a dozen or so of his more lusty friends: players, half-assed players, and hangers-on, many of whom, as the cliché goes, would screw a snake if someone would hold its head.

"Where did he get the name Jaguar anyway?" one of them asked of his friend Dirk, one of the flunkies who always seemed to be at Jaguar's elbow. "He drives a Rover."

"Probably because he tore up more young pussy than any player in the sixteen-and-unders when he was the great white hope of the LTA," Dirk replied. "Anyway, mate, he moves like a bloody cat. Haven't you ever noticed?"

They turned back to Jaguar, who was speaking to the group in general. "You know he went broke and died a pauper, Hansom did," Jaguar said, turning his dark face upward. "The city took his patent and never paid him a farthing. Now, I think that stinks! Let's lift a jar to Hansom. Hansom is as handsome does, and all that shit."

The men and their female camp-followers hoisted pints to the gently swaying wagon and drank solemnly to the diddling its inventor received at the hands of the establishment.

"Time, ladies and gentlemen, time please. Drink up and clear out, please. You don't have to go home, but you can't stay here," the governor bellowed, ringing the bell. Dirk suggested that they go on to Miranda's or to his flat.

"I've a couple of very good bottles of Beaujolais back at the flat. Why don't we push off and have a party before you have to get serious about doing your thing next week."

"No. Not this time, thanks," Jaguar said. "I'm having a workout and a knock with Billy Sherman early in the morning. I don't want Beaujolais bubbling round inside my head when we're running the track in combat boots."

Jaguar bore a strong resemblance to Roger Taylor, another former England number one. They both were almost Australian-like in their religious insistence on physical fitness. Both had dark hair and Celtic features, but Jaguar's eyes were so pale blue that it gave one the impression of being able to look right through his head. Women were devastated by them. He was, consequently, able to play the strong, silent type, which was a damn good thing since he was quite inarticulate and, for a superb athlete, strangely clumsy off court. Yes, he moved like a cat, but a very clumsy cat.

"Don't work so hard, Jag. Death is nature's way of telling us to slow down," Dirk persisted.

No, Jaguar didn't possess Taylor's poise or left-handed swing serve, but he did have the Yorkshireman's tenacity of purpose. Bumping his knee on the bottom of the table, he stood up amid a symphony of rattled glasses.

"Aw, come on, Jago," the disappointed chorus of groupies wailed. "Just for an hour or so—we won't touch your precious bod—you'll breeze Queens anyway," they chirped all together. The allusion to the upcoming Queen's Club tournament was enough to fill his sails with wind.

"Time for noddies, my darlings. If I come a cropper in the early rounds, I'll screw the lot of you." He waved his good-byes and prowled out into the English summer night.

In another pub in lower Earl's Court Road, the Australian ghetto of London, a similar scene was being enacted. Made in the same mold which produced the greatness of Laver and the other Harry Hopman greats, the young crop of Australians—confident, conditioned, maintaining the Australian tradition of power tennis—stood near the bar sucking up pints of Foster beer. They were as nervous as anyone else about the projected difficulties of Wimbledon, but their nervous energy was channeled through a transformer which turned it into a rapacious will to win.

This new breed differed from their Hopman forerunners only in that their faces and styles were personalized. The only American among them was, understandably, an agent.

"A man's house is his hassle," Jesse Fraser said, blowing the foam off his beer. A bit of it flew onto the table, and Jesse covered it with a round coaster.

"Are you sorry about getting married? I mean, Christ, mate, she's fantastic!" Wicko said, meaning it.

"Oh no!" Jesse corrected, sing-songing on in his heavy Queensland accent. "Marriage beats being blind and crippled any time. I tell you, swaggie, she just sends me 'round the bend before every big match. Like she intended to or something. Chroist! I'm nervous enough without all that. Hey, bartender, let's have another round of frosties over here!"

"Ho! Hold on. It's my shout." Al Wick put the knuckles of both hands on the table and his massive shoulders propelled him into a semi-standing position. "I'll take them here, mate." He moved over to the bar. "What d'way owe you?" The bartender pulled four pints, saying, "You've been in here all night and the price hasn't gone up yet, so I figure it's about the same as last time."

"All these Pommies are wise-asses. Can't we just get served without any of your shit-ass remarks?" Jesse leaned between Wick and the bar, ostensibly to help with the beer but actually to cool any escalation.

"Don't get hot, Wicko. Save your steam for the fifth sets. You'll need it. He's not English anyhow. He's from Alice Springs."

Wick was still glowering at the barman as he turned to move off with his big fists full of the best of bitter.

"Fucking hell! Bloody fucking hell! What a bloody nerve, right?" The eight pints he'd already had weren't sitting at all well on a dismal, erratic practice session. Jet lag and nerves combined with a slight travel cold to put his timing off and shake his self-confidence. This, just before a major tournament, was rather hellish; before Wimbledon, it was the trigger for a devastating depression.

Phil Katz, the American agent, was talking to Jesse, one of his most lucrative clients. "If your wife is getting to you like this, something has to be done. If I were you, I'd put her through some heavy head changes."

"I'm not all that sure she's not right," Jesse said. "I mean, what are we doing beating our brains out in the hot sun, chasing a little yellow ball about—in our underwear—churning our stomachs and rasping our nerves raw week after week?"

"Making pots of money, that's what, Digger," Freddie Moore interjected sarcastically. "Now's the time to grab the lolly before the chance slips by. Shit, you know that the nerves go before the legs do, and that's the name of the game. At our stage—like—pretty even in stroke-making and all, it's concentration, confidence what makes the difference. Confidence ain't nothin' but nerve. You know. Psychology and that sort of rubbish."

This outburst exhausted Freddie and he returned seriously to his beer. He looked into the mirror but didn't see the foam on his lip. He was gazing through the glass into a future of confidence hitting through the ball.

Ten to eleven showed on the pub clock, always ten minutes fast. The publican gave last call. Wick asked what the chances of getting an indoor·court at Queen's were. Jesse had one booked, and they arranged to hit threes, two against one rotating, on Jesse's time.

"Wicko," Freddie asked, as they piled out through the swinging doors, "are you still going round with Missy Teaford? Your circuits go different ways, don't they?"

"Who, her? No! That douche bag—I don't see her no more. If she had as many pricks sticking out of her as she's had stuck in 'er, she'd look like a flaming porcupine."

"Charming," Jesse said over his shoulder, cramming his long, wiry body into the Triumph. "Sounds like a man scorned to me, mate. Sour grapes?" He was tooling off down the street before Al Wick could arrange his sputtering protestations satisfactorily.

Phil Katz hailed a cab and bundled a blond groupie into it. "The Gloucester, please."

Wicko, Freddie Moore, and Paul Bell, a doubles specialist on their Davis Cup team, started off down Earl's Court Road to the flat they had rented for the duration of the two tournaments. They would all be long asleep when a fourth flatmate crept in later with one of the better-known girl players.

2

"Wake up, Francis," Shep's father whispered in his ear. "Happy birthday, buddy! Twelve-year-old guys don't lie around in bed at seven o'clock in the morning. You have a couple of hundred balls to hit before breakfast."

"Oh, Dad. Even on my birthday?"

"You don't have to if you don't want to, but I'm not the one who has to play the finals tomorrow. It was your idea, pal. This is the one day I can sleep in. So what's it to be? I'll jump back in my warm bed."

Shep labored out of his dream-warm cocoon and stood in a daze next to his father. They were the same height, which made the short, stocky Mr. Shepard and his small, delicate-boned wife marvel. Proteins or gamma rays on marigolds or whatever, today's youth were shooting up past their parents.

The cold California morning mist glided and swirled down the hills and engulfed their modest cliff home. The sun didn't burn it off until ten and sometimes eleven o'clock, and there was chilly damp permeating the place. Shep doused cold water on his face, picked up his toothbrush, thought about skipping it, then, with a supreme rush of discipline, brushed them. He was proud he'd overcome one of the first obstacles, one of the tests of his innate laziness. He put off making his bed.

Shep's father, Michael Shepard, was a history teacher and tennis coach at the day school. He'd never done well at tennis himself. His strokes were crude and inadequate, and his less than average height gave him a handicap that his superb concept of the game couldn't hope to overcome. His own father, Shep's grandfather, was a musician-turned-carpenter and consequently hadn't achieved country-club income or status. And at that time, country clubs were where it was at as far as tennis was concerned. Shep's father was largely self-taught, very keen on the game; in fact, it was his passion.

Although he made it amply clear to Shep that they were father and son, he never addressed him directly as such but rather in the parlance of friends. As he brought Shep along in tennis (and other things), it was more in the manner of an older brother or friend. He tried within reason never to force him. This was sometimes painful to the elder Shepard, because the boy was fairly exploding with natural talent; yet he was also probably the laziest child on the West Coast. Michael Shepard had to wait until the boy was moved by his personal muse into action. This happened a good deal less often than it should have, but still Michael refused to push him.

The surf pounded below, agonizing the cliffs and spraying a light veil over the picture window of their combination breakfast—sitting room. Mrs. Shepard taught science in the high school. Her school holidays and breaks often coincided with those of her husband, but not this particular one; she had already left for classes, and the men were thrown on their own resources in the kitchen. Breakfast was less than your average *succès fou*.

They drove through the fog up the valley to the school courts.

"I'll give you your present tonight when Mom comes home, Shep. She had a big part in choosing it, and I think you'll agree that it's only fair to wait."

"Sure, Dad. That's cool. God, it sure is wet-looking this morning. Those balls are sure going to be heavy. I hope I don't get a sore arm or tennis elbow out of it."

"Don't be such an old lady. I'll pick out the most worn balls. That should even things out. Just put negative thoughts out of your head and concentrate on hitting every ball right."

As they walked to the court carrying their gear and two baskets of balls, Mike Shepard looked at his son, now ambivalently bursting with anticipation and fear about his upcoming final in the twelve-and-under state championship. His eager face reflected a coming generation in the game, and Mike Shepard wondered about their future. He was a maverick LTA member who believed desperately in change from within. The Lawn Tennis Associations over the country did remarkable work at the lower levels. Their dedication and sacrifice were really impressive. At the higher echelons, though, greed-power plays sullied and saddened the full operation, and Mike Shepard thought of quitting and going it alone. Margaret Court had been forced to such a step in Australia years before and had come out smelling of roses. It seemed to Mike that Lamar Hunt was the only voice of sanity; opening up the game to the world rather than allowing it to stagnate in the plush, lucre-laden clubs, country and otherwise. Hunt was eminently fair in his treatment of the players in his group. For the first two years he took a terrible beating in the pocketbook, before television disseminated the game's combination of elegance and stamina to the public. In dealing with the LTAs of the world, Hunt's voice was usually that of reason, and when it wasn't heard over some of the more suave accents, he finally took off on his own tack. Inevitably, they came back to him with "solutions" (usually those prescribed by Hunt in the first place) and they came to an accommodation. But Mike Shepard felt that it was too late for the LTAs. They hadn't seen *"Mene, mene, takel upharsin"* on the wall soon enough. Their inordinate, clubby, amateurish power was slipping through their reactionary fingers. Young Shep was still under their aegis and was well treated. But in the years to come . . . ?

Without removing their warm-up suits, father and son began hitting, gently at first, careful not to strain anything. For the first twenty minutes they hit forehands down-the-line, using only the right half of the court; then backhand cross-courts for fifteen minutes.

"Keep the head of your racquet up on those backhands, Francis! Concentrate! Move your feet!"

Shep's feet had grown faster than the rest of his body, and he was suffering through the necessary adjustments.

"Okay. Anticipate, Francis, I'm going to spray them around. There's no shortcut to anticipating. You won't get it from watching my feet or anything weird like that. Watch the ball come off my racquet and prepare immediately. All right . . ."

They rallied full court for a few minutes. "No! You're late. Get that racquet back sooner . . . move with it back. All right!" he exclaimed with obvious pleasure. "That's better. Now don't try to hit so hard. Step into the ball smoothly, with your weight forward. Don't jerk it."

Shep was gliding into the ball as well as his big feet allowed, and except for the occasional lapses normal to the attention span of a twelve-year-old, his loop-swing was flawless.

"Change the pace, change the spin, Francis. Don't get into a groove. Think! Now come up on a short ball. Follow the line of your shot up to the net."

"I know that, Dad, damn it!"

"Don't swear at me, pal. I'll rap you one. You can't hear it too often. We're programming. Now, hit that approach shot deep."

Shep came in on a short ball and tried a backhand chip to the corner. It flew out.

"The first rule of an approach shot is that it must go in. Let's try again and . . . keep it smooth!"

Shep hit a topspin drive down the line and moved up behind it, volleying well out in front, but with too much swing. It sailed against the fence.

"Right, killer. Now tell me why you boobed that shot."

"Aw, Dad. You know! I took too much of a swing at it."

"Don't take any. Don't think 'swing': punch it, and let the racquet do the work. And don't try to make the first one a winner unless it's a sitter. You're not Dennis Ralston. You're a kid. Hit it deep to the corner and . . . what?"

"Follow the line of the shot for the second volley and angle it," Shep said in litany, exaggerating his father's presentation.

"Don't get sarcastic. Just do it. If you'd prefer to go play baseball, just say so."

"Come on, Dad! I didn't mean anything. It's just that you've said 'don't' eight times in the last ten minutes. Can I work on my serve for a while?"

"In a few minutes. Let's try to get this volley straightened out. Come to a stop—on your toes—and punch it. Get down to it if it's low. Also, Francis, break your wrist a bit more. There, that was better. Don't look where you're hitting it. Watch the ball, and no matter how well you hit it, *cover!* Cover the return!"

They practiced the volley for several minutes. Then Shep served a basket of balls while his father made a phone call. When he returned, they did two laps around the arena and piled into the car. Shepard spoke as he drove. "Well, Francis, I don't know what to say because the boy you're playing is an unknown quantity. The first seed had to scratch his quarter-final, and this young fellow pulled an upset in the semis. I don't want you to be nervous because you've never played him before. But you can't get overconfident either."

"I think he just lucked into it. Charley killed him last month up north. He hit a lot of ass shots yesterday."

"There isn't any need for that sort of talk, Francis."

"Why, Dad? You say it sometimes."

"I guess you're right. It is a little hypocritical, but it doesn't sound right coming from young people. I know you all talk like that when you get together . . . and I suppose we should be honest about it. But let's keep it to a minimum, okay? Constant swearing dulls the mind and promotes a very lazy vocabulary. As long as we're on the subject, let me tell you this quick story. The point of it is your volley. You are still trying to hit it on the run. There are these two bulls up on top of a hill, you see. The young bull says to his father, 'Let's run down the hill and screw one of those cows!' The older bull replies, 'No, son. Let's walk down and screw them all.' You may make a few of your volleys on the run, but your percentages will go way up if you slow down, hesitate, and volley with firm balance."

Shep was glowing in the realization that his father had talked to him man to man for the first time. He felt a flood tide of inner warmth for the rest of the ride home.

Shep had a wonderful evening. His parents and ten of his school friends surprised him with a funny, larky party, and he was on a magic cloud most of the night—which terminated with a small display of *Son et Lumière* after the others had left. The heavy, ionized weather broke dramatically with a stupendous show. The night was almost apocalyptic; an electric storm split the blistering heat with a bristling explosion—a whiplash of jagged light followed by the roar of a billion dead souls. (This last, a wide-eyed interpretation by Shep as he watched the action out over the ocean. His intense excitement was a mixture of genuine fear and the morbid delight of his imagination.)

As the storm passed, he succumbed to reality and began to wonder about the courts the next day. They were concrete courts, and the match wasn't until two-thirty in the afternoon. So, knowing they'd surely be dry enough, he tuned down into a deep but twitchy sleep.

Even in the warm-up, it became very apparent to Mike Shepard that his son—right on top of his game—would have no trouble with the other boy. When Shep won the first set 6-1, even easier than the score indicated, his father turned to a friend, Dick Nutter, also an active LTA member, and became deeply engrossed with him in a conversation on tennis politics. They drifted off to the club bar. The Association of Tennis Professionals was the contentious point. Mike Shepard thought that perhaps, just perhaps, the ATP was now making noises and moves to supplant the ILTF as a power force, *the* force in the game, and was already undergoing the hardening of the arteries expected from a much older organization.

"I think they've done a great job. I mean—facing down the Wimbledon creeps and getting all those guys to hang in there together," Dick Nutter was asserting over his second scotch.

"In the first place, although I agree that they were wrong to have —to cause—a confrontation at and during Wimbledon, I take offense at you calling those fine men 'creeps.' If they did make a mistake in judgment with ATP—and I personally believe they did— they also were responsible for open tennis in the first place. Those 'English creeps,' or however you put it, were the people with enough courage to take on the entire International LTA. Go it alone—and win!"

"Sure! I know all that. I thought you were the one who was down on the LTA. What about that?"

"I'm not against any entire group of people in any segment of life. In the case of the ATP and the Wimbledon committee, they both misunderstood and underestimated each other. Say, Dick? How long have we been in here?"

"About a half hour or more. Why?"

"That match should be over. Let's see what the hell is keeping Francis. He always gets embarrassed if I'm hanging around to congratulate him if he wins. He should be finished long ago."

The two friends walked back to number one court where, to the surprise of both of them, the score was only 3-1 to Shep. Instead of putting volleys or setups away, Shep was showboating all over the court—hitting tantalizing touch shots and teasing the other boy with lobs when the point should have been over. The disparity in their level of play was painful, and the mother of the other boy looked close to tears. During the change at 4-1, Shepard called to his son, "Are you trying out there, Francis?"

Shep looked a trifle hangdog through his stage smile. "Sure, Dad."

Mike Shepard walked out onto the court with as much dignity as his short body could muster in his embarrassment, grabbed Shep by the ear, and hauled him over to the umpire's stand.

"We respectfully default, Umpire."

He took Shep off the court and away from his first big tournament victory, impressing a lesson on Shep he never forgot, one of humility, tough play, and sensitivity. It was many years before Shep understood and forgave

3

The next four years, as so often with children and adolescents, were an eternity for Shep: one-fourth of his lifetime, and his days were commensurately filled and charged. Besides his tennis (which waxed and waned in interest), he played varsity basketball and even played end on the football team until the teachers complained about his grades. His father variously cajoled, explained, and argued about the need for concentration of effort. Finally he had a secret talk with the principal. Shep was manipulated and the football team lost its tallest end. Two weeks later Shep joined the boxing squad, and Mike Shepard threw up his hands, threw in the towel, the cards, and the shooting match.

Shep's marks improved, though, and at just the opportune time, he found interest and form in tennis. He was big and strong. Although his feet were still too large, he moved well. His ranking improved until, at the optimum moment, he was in a tie for number one in his area. The regionals were going to be the decider and, in all probability, a brash young boy from L.A. would be his opponent.

Randy Mariano, even at sixteen, was renowned as a hustler, bad sport, and wise guy. Even in "friendlies," the boy playing him had to find a buddy and press him into service as an arbiter. Randy was also brilliant. His game was sound, his tactics uncanny. All of his cheating—the dodgy line calls and forgetfulness about the score—were merely cheap shortcuts. His game was good enough without them, which made it all the more shocking and sad.

Shep had a close call in the first round, but otherwise reached the finals in straight sets. Mariano breezed through it too.

So the stage was set and the players groomed and rehearsed. Randy jumped off to a 3-0 lead with a quick break of serve. 3-0 and 5-2 are the most deceptive scores in tennis if there is only one service break. Mariano went on to a 5-2 lead, and Shep broke him back for 5-4. He held his serve at love, volleying well off the low chip returns Mariano was cultivating. At 5-all, Randy had a dubious call at 15-love and threw a fit of such obviously bad acting that instead of putting Shep's concentration off as intended, it made him laugh. The laughter completely undid Randy. A chink in the armor —it cut through his cynical defense mechanisms. Randy could not stand to be laughed at. He was totally flapped, dropped his service, and Shep ran out the set. As is very often the case with young, inexperienced players, Shep had the "first-set letdown syndrome," and Mariano got on top again. He won the second set, 6-4, and they had a fifteen-minute break. In the locker room, Shep yanked off his clothes and hopped into the shower. Mariano stretched out on a bench and began his hustle.

"I've got you now, you big skinny jerk. You're falling apart," Randy taunted. Shep started singing loudly in the shower (to the tune of "Everybody Loves a Baby"):

"When you're oversexed and unrelaxed
And don't know what to do,
Have a baby, have a baby . . ."

Mariano jumped up from the bench. "Look, you big cunt—can you hear me? I want to be number one. How much will you take? Being two isn't so bad. It won't kill you—but I've got a big bet."

Shep turned the shower off but didn't come out.

"Will you take two hundred dollars?" No answer from the shower. "Three hundred?"

Shep said a firm "Okay."

Mariano started putting on a dry shirt, saying as he pulled it over his head, "I'll give it to you right after the match."

Shep stepped out of the shower and toweled off. "Now."

"Now? Where will I get it now?"

"The same place you'd get it after the match," Shep said, pulling on his jock. Randy ran out of the locker and Shep continued dressing. A few minutes later, Randy barged back in and slapped the money down on Shep's racquets. Shep slipped it into the racquet cover just as the referee rushed in.

"What the hell's keeping you guys? You're running way over fifteen minutes. I've a good mind to disqualify you both."

"Not likely, man," Randy snickered. "But it would be far out."

For Mariano the third set was a nightmare. Relaxed by his "purchase," he was totally unprepared for the blitzkrieg Shep unleased on him. Twenty-five minutes later, at 6-1, it was all over. Shep was first into the locker room, followed closely by a hot, screaming Mariano.

"What the fuck did you do that for? Jesus Christ!" He smashed his racquet into pieces against a locker. "You sonofabitch! Give me back my money!"

"Sure," Shep said calmly. "Here." He placed it on the bench.

"What—why did you—what the shit did you do that for?" Randy sputtered.

"Oh, I never meant to lose, Randy. But if by some misbegotten chance I did, I had three hundred dollars. That's a lot of gas, beer, and pizzas."

Without showering or changing, Mariano grabbed up his belongings and stomped out of the room.

4

At the end of the season, when the national computations were all in, they showed Francis Shepard to be eligible to represent the United States at Junior Wimbledon. There were still some big tournaments in the States that the LTA wanted him to enter as their "boy," but it was inconceivable to Mike Shepard that they would force the issue—deprive young Shep just for their own ego. They usually held out the heavy iron to swing in cases where pressure was needed in their major tournaments: who would play and what cities would host them. The Davis Cup was their special domain, the big plum, and the politics, both internal and international, was often not a pretty affair. Mike Shepard had worked within the organization to protect the individuality of the players. As champion of their rights, his success was unspectacular.

It was decided to have an informal meeting, a weekend of tennis at Palm Springs, to discuss everything, with some swimming, a social round robin, and a party for Shep in gratitude for his hard work and accomplishments. Intuitively, Mike Shepard considered begging out of it. It would be stuffy. All the self-important hierarchy with their wives, and all the hang-ups the flimsy position of dilettantism entails. THE ULTIMATE YES WORLD IN THE BEST OF ALL ULTIMATE YES WORLDS. High-blown egos in the make-believe world of sports.

There was something that didn't ring quite true in the conversations that bothered Mike Shepard. No. Surely they couldn't, wouldn't sacrifice Shep's chance to go to England just to play in their tournaments. He relaxed, knowing finally that it was unthinkable, realizing that his growing prejudice had impaired his judgment and fairness.

The first round of cocktail parties bogged down in heavy drinking and hors d'oeuvres—gorging at private homes. Shep, of course, wasn't invited or expected to attend, and his disappointment was registered all over his big happy smiles.

The dinner, which he did have to make, was a study in back-slapping and self-serving after-dinner speeches which were embarrassing in their lightweight boy-scout banter. Grown men grown silly on bourbon and martinis playing ping-pong with compliments in long-winded mutual admiration discourse; a maze of words threading a path through glorious self-denial and sacrifice, high-mindedness, getting a bit lost, but, yes, the light at the end of the tunnel: the speaker had finally found his way out of his own mental morass:

". . . and it was all worth it. We came up with a boy, a hard worker, a talent, a boy with promise. Francis Shepard!"

Applause!

Shep was sound asleep. After a hard morning's practice under the hot desert sun and the inane round robin with the LTA members and wives, he had spent the remainder of the afternoon trying to screw a long-legged young thing in a cabana as ineffectually as only a male virgin can. Hours of "almost" . . . hot and cold . . . "yes! yes!"—then "no! no! no!"

His every fiber was exhausted, and the speeches pulled the plug on his last reserves of energy.

Mike Shepard nudged him and whispered in his ear. Groggily, Shep lurched to his feet and waved a glassy-eyed "Thanks!" to the other tables.

When it was over, the vice-presidents took Mike Shepard's elbow and, with an arm around his shoulders, steered him aside.

"We'll have that chat tomorrow morning by the pool. With the Bloody Marys. Right now I'm just bushed. Good night, Mike."

Under the bright blue, airless sky, the desert lay shimmering in its calm beauty. Shep joined the group at poolside, shook hands courteously, and sat down in an unobtrusive place outside the perimeter of elders. His father was the object of attention, and the expression on his face indicated a strong possibility toward a swing to displeasure any moment. The LTA president was beefy and slope-shouldered, and had a permanently furrowed brow from disagreeing with everything. When he spoke, the words popped from his little round mouth like gas bubbles from swamp mud.

". . . and so you see, Mike, Francis is right on top of his game now, playing on cement. The two weeks in England on grass—he'd probably be lost on it anyway. He's never played on it before, has he? Has he played on it before?"

Mike Shepard shook his head, not bothering to add that ninety per cent of the other boys had never played on it either.

"Right! So I'm sure you understand and agree that it would just ruin him. He'd have to get his game together all over again when he came back."

"That should be my decision."

"Well, of course," the president said, putting down his cigar and picking up his gin and tonic. He toyed with the lime for a moment and continued. "But you see, it's our job, our duty and responsibility, to guide you. I mean, we've got a lot invested in the boy, and people want to see him in these tournaments. And afterwards, too. You know, college and everything. We can be very, very helpful there."

"Whether or not Francis stays under your auspices is up to him," Mike Shepard said quietly. "We'll have to discuss it. It will be his decision. I hope my action here today won't hurt him later, but I feel I must take it. It's long overdue."

"What action?" the president said.

"I resign," Mike Shepard said. "I leave you to drown in your own bullshit."

He shoved his metal chair back with a grating rasp and wound his way among the tables to the door. The rest of the table stared stunned in total incoherence. Shep, in complete confusion, hesitated, then followed him to their rooms.

"Wow, Dad!"

Shepard was putting his clothes into a suitcase on the bed. Shep sat down beside the case and watched with that strange, riveted attention given to ordinary actions during moments of uncertainty and crisis.

"Only two of those parochial fools have ever been to England—only one of them to Wimbledon," Mike Shepard said, more to himself than to his son. "I had more than an inkling that the theme of the 'chat' this morning was going to hit along these lines. I struggled all night with whatever my line of action, if any, was going to be. You see, Francis, I've always wanted, dreamed, of playing Wimbledon; even acted out some of my fantasies. I don't consider it foolish. It was fun. I enjoyed it. But I didn't want to vicariously fulfill my little whimsies at your expense. I know the tournaments coming up here are fairly big potatoes and you have many years ahead to go to Wimbledon. Yes," he shuddered as he said the next, "one never knows what quirk of kismet, what disaster, even, is lurking around the corner. I've never believed in *carpe diem*—'take the cash and let the credit go,' that sort of philosophy—for myself. I should never have made a teacher if I did. My life has always been an endless vista of living and teaching. You are different. I want you to grab your chances. I'm not unhappy with my lot. Just numb. For you, life will be another thing entirely. Those greedy bastards. The president said the operative word: 'Investment.' They weren't giving of themselves for the sake of tennis. They were 'investing.' They're a bunch of spoilers right out of the system, the big boys of the LTA. Do you know what he said to me? 'Grass is for grazing.' Can you believe that? And you know what else? 'Wimbledon will probably be converted to cement like the rest of the world.' His little world, he meant. Can you imagine cement at Wimbledon?"

Shep didn't answer the rhetorical question, but pulled a face in feigned corroboration. He couldn't imagine grass!

"Well, that's it for me, Francis. I've put in a good many years with them, and some very rewarding things have come out of them. But in total, it should have been much more."

"Everything should be better, Dad," Shep contributed, precociously.

"Now ain't that the truth! You think about it and come to your own decision about staying with them. I can't afford to send you to London and they wouldn't sanction you anyway, so it's moot. You better stick with them, I guess. They have all the cards. The players are pawns to them. They mean as well as they can in their own selfish, pompous way. Pack up and let's get going, Francis. We've a long drive—a long way to go."

5

The first time it happened—the first time he really understood—
impacted the greatest shock of his young life, and it began as the
most mundane of possibilities: the act of petting a dog. It was
Shep's seventeenth year. Because of a quirk in the timing of his
birthday, he was already a freshman in college, already in the whirl
of fraternity rushes, hot-eyed little coeds, heavy schoolwork, and
the tennis varsity.

The fresh-mown grass glistened with baubles of dew in the early
morning sun, which was building its strength to heat the day. Again
the natural, grassy aroma wafted up from the huge rolling lawn
that stretched from pseudo-gothic to pseudo-gothic.

He hadn't seen the dog coming. It had the effect of materializing
from another time continuum. It was just there. Shep spread the

morning paper he'd finished at breakfast on the damp grass and sat down, his upturned face gathering rays and his long fingers scratching love-energy behind the dog's very willing ears. He thought, "Let's enjoy this moment, old boy. It's fine." Then, aloud, "Feels good, right? I guess all the bad rubbish that happens to us heightens the good experiences. Well, enjoy it! Inhale it, you good old guy. You've got a very short little life and you'd better savor it."

Like a cold, electric chill that permeated his entire body, his being, it riveted him: the realization, the crushing, imperative realization of his own death. Not the theory of it; not the abstract, unbelieved knowledge of a silly certainty, but the actual, all-encompassing horror of the real inevitability. Himself, no longer. The world empty of him. And *when? How?* Perhaps not of old age in the seemingly endless vista of days. But by mischance; misadventure. The fear swelled through him, sucking energy with it, leaving him hollow and wide-eyed.

Philosophically, life had more and different meanings for him. "What the hell" was the first reaction, and it showed in his studies and everything else he did. All appeared ridiculous and irrelevant. (It was to be several years before he was to channel it into the theory of the absurd of existentialism.)

His tennis coach walked with him to the locker rooms after one particularly exasperating and dismal practice session.

"What's your bag, Shep? What's got into you? Or should I put it, what's gone out of you?"

"What do you mean by that, coach?"

"Don't give me that shit, kid. I know you're bright, so don't waste both my time and yours pulling a dumb act. You're not working. Not working and not trying. You can't come on my squad and just walk through the motions."

"You mean go through the motions or walk through the play."

"Don't be a smart-ass! You know perfectly well what I mean. What exactly is your problem?"

"Well, it's all so silly, isn't it?" Shep stammered. "I mean, for Christ sake, what's the point? There must be, there's got to be something more important to do with the short time we have in this world. What the fuck's the big deal whether I can hit a tennis ball harder and more often than some other idiot? Who's going to know or care in a hundred years whether I was number one on the team or beat Benny or Randy Shitface? What in the world is the difference?"

"I'll tell you what's the difference, Plato. I'm dropping you to number four. If you don't get the lead out of your ass by next Sunday, I'll drop you completely! I don't care what the LTA or alumni says. Understand? Get straight, kid. Are you smoking dope?"

"No, for Christ sake!"

"Grab a shower, I'll talk to your counselor about it. See you."

Shep daydreamed through three classes. He left the campus for the parking lot. As usual, his second-hand Pontiac took five minutes to turn over before starting, and he interpreted this as part of the new conspiracy against him. The lack of a positive force and direction had swung him to the obvious: a cynical and slightly paranoid state.

The drive to the public courts to pick up Missy was somewhat uplifting, though; a salt-edged breeze fanned into his nostrils and ears and softly touseled his hair. He found himself grasping at these odd moments of exhilaration as if trying to array enough of them into a mosaic of happiness. Trying to gather enough of them to create some substance, some semblance of life-style.

Missy Teaford wasn't a beauty. She was handsome and fresh-looking. Her firm, round, contoured behind was her best feature, presenting as it did what seemed to be natural hand grips. Missy was a salvation to him because she was low key, sweet, and astonishingly perceptive. She understood. As a player herself, she understood the need for tennis in his life, and—although on the level of teen-agers —she loved him, blindly and thoroughly. Missy shared his growing pains and tolerated his aberrations.

Shep swerved around a Good Humor truck and parked behind a cluster of racked bicycles. Having cut a class, he was ten minutes early. He left the engine idling while he dashed over to the court where she was hitting. Missy's honey-tanned body darted back and forth on long, well-made legs. He watched the rolling of her ripe, sensuous rump as she walked back to the fence to pick up the balls. Aching desire caught at his throat. When Missy spotted him, her dazzling teeth signaled a spontaneous reaction of good feeling; a genuine "Glad to see you!" Clowning, she flicked her eyelashes exaggeratedly, and gave a coquettish little wave with wriggly fingers.

"Hurry up, my engine's running." He smiled shyly to himself about that. It sounded like a confession.

"Be right there!"

Shep picked up three stones and began to juggle them. He dropped two of them, and, as he chucked the third at a tree, he noticed the policeman giving him a ticket. His lope back to the car was a study in phoney nonchalance.

"This your car?" The cop didn't look up from his writing.

"Sure. What's the bust?"

"Big talk. You left it running and unattended. Smells like it's gonna blow up."

"How much is that going to cost me?"

"Ten bucks. You can send it in. Will you shut that thing off?"

"For ten bucks, you turn it off." In a WHOOSH of hot spray, the rusty cap zonked up inside the hood. Shep turned eighty degrees, shoved his hands in his pockets, and feigned interest in the trees. At times like these he wished that he smoked. The sophisticated gesture of lighting up would have given him just the correct posture to finesse the moment.

The cop walked off chuckling as the engine geysered over and flooded the gutter, planing popsicle sticks along like tiny surfboards. Shep watched them with negative interest.

Missy came padding across the grass on silent sneakers. Only then did he reach inside and switch off the engine. She had stopped and was disconnecting a hose from a sprinkler, but she had to leap back every two twists as it chug-chug-chugged around in her direction. She had changed her shorts for clinging silk slacks and didn't want them spotted.

"You're getting pretty scientific, princess," he called across the lawn.

"I've had enough practice. Why don't you shut that darn thing off when you come over here?"

"You have a beautiful ass!"

"I really wish you wouldn't talk like that, Shep." There was an obvious ambivalence in her voice.

The water replaced, they burned rubber off down the boulevard, which was gee-gawed with imported palm trees. The trees were a shock of incongruity against the gothic buildings; one never became inured to them. Every day—same jarring assault on the senses.

"Where are we going this afternoon?"

"Come on, Missy. Where's your sense of adventure?"

"I don't want to go too far."

He winced at that. "I've got a piano recital tonight and an exam to prepare for Tuesday."

"I thought we'd buzz down to my dad's beach shack and swim—and like, you know—talk and fool around a bit."

"That's a half-hour drive."

"Twenty minutes. You aren't scared are you? I mean, you still trust me?"

"I'm the one I don't trust. I know virginity is old-fashioned, but, well, I'm still very up-tight about it."

"Tight for sure."

"Don't be crude, Shep. I'm serious. It's looked at by my girl friends as a deformity."

Shep slowed down and cautiously merged onto the freeway before gunning it again.

"One of my favorite—I should say my dad's favorite—stories is just like us. Two college kids who live in dorms with no place to go to make love, and she's not going for it anyway. She's a virgin, too."

"Is this a dirty story, Shep?"

"No! Just kind of a cute story my father tells. Anyway, it's November back East. You know, cold and everything, and they're wheeling around one night in his little MG. The old, funny convertible ones."

Shep pulled out into the passing lane to overtake a semi, then angled back to the inside lane.

"Anyway. He's giving her all the seduction pitches. You know—like the psychological one: 'You're repressing it, and it's going to be a bad seed in your psyche, in your unconscious, and, and give you a trauma later.' Then he drove up into one of those old New England cemeteries and he's giving her the philosophical shot. 'See all of those dead people? Life is very short. No one knows better than they do. Why, they would trade their entire lifetime for a little sex, and here you are wasting precious time.' Then she says finally, 'All right, already. That's enough. I give up. Let's do it!' Well. They can't do it in the little car and the grass is all cold and dewy wet. So they decide to do it on one of those inlaid tombstones.

"Afterwards . . ." Shep glanced at Missy, who was nodding and smiling appropriately but with caution ". . . afterwards she went back to her dorm. A little stiff and hurting a very little bit, but not as changed as she thought she would be. Anyway, she decided to think about it for a while and not tell her roommate, who was a real bitch. Her roommate asked her, 'Did you have a good time?' because she looked rumpled. The girl answered, 'No, and I'm never going out with Charlie again. He's like an octopus. Hands, hands, hands! He's everywhere.'

"Well, she was taking her clothes off as she said all that and she asked her roommate, 'Am I covered with bruises? I probably have bruises all over me.'

" 'No bruises,' her roommate said, blowing a ring of smoke like Bette Davis, 'but your ass has been dead since 1881!' "

Missy exploded with genuine laughter and giggled until tears rolled down her cheeks, partly from the joke and partly from relief. She had almost made up her own mind.

"Out of sight, Shep. Really. That story is far out." She chuckled to herself from time to time through the next five minutes it took to reach the beach house.

The shack (for, from the outside, it did indeed appear to be one) nestled solidly on a large ledge about three-quarters of the way down. Its weather-stained clapboard shingles were once again the original wood color, and looked more Cape Cod than California.

"It looks crummy but nice," Missy said. "I'm glad to be here. I'm glad to be anywhere, after those steps."

"Really!"

Shep fumbled through his bunch of keys, searching.

"If you can't find the key after all that, I'm going to shove you off."

"Got it!" he exulted.

She gave him a victory hug that lingered into a feverish, hot embrace and the timeless magic music of the surf below added its erotic tempo.

Shep's hand trembled slightly as he jiggled the key into the lock. He was unconsciously humming "Here Comes the Bride." Inside, the shack was transformed into the coziest retreat Missy had ever entered. Heavy wood paneling on two ends—one end a combination bookcase and fireplace; on the other end, a huge old ship's steering wheel was mounted with a framed drawing of the original frigate below it. The sides of the room were covered with rich brown felt, and the chairs and couch were black leather. It was a very private place. Shep had always disciplined himself against guessing why his father created it.

The heat from the embrace outside was little diffused as they walked hand in hand over to the picture window, spattered white-opaque by the salt spray. Not much light filtered through. Shep leading the way, they climbed down the metal spiral stairway to the kitchenette and bedroom below. He opened the tiny refrigerator on a treasure hunt and withdrew two fine cans of beer from an intact six-pack.

The bedroom was very simple, lackluster in fact with its plain wrought iron bed, white bedspread, and a highboy dresser peeling its white paint. Shep's mind was racing. No. He discounted the bedroom in favor of the long white shag rug upstairs. The bedroom was too stark and unromantic, and, goddamn it, the first time for both of them was going to be romantic. That is, if she didn't change her mind again. He had convinced her to start on the pill two months before (at least he felt reasonably sure he'd convinced her) and still . . . nothing. He was burning up now, and his efforts at nonchalance were cloying in his throat. When in doubt he pulled on the beer. It was a novel experience, given his recent negative frame of mind, to be excited about anything.

"Let's go back upstairs. It's more comfortable." He took her by the hand and they wound their way up. In the living room, he offered her the beer but she silently declined, her eyes hot on his face.

Shep pulled her toward him and embraced her more roughly than he fully intended, as though afraid that she might vanish, or at least change her mind. (He was quite sure that at some point that afternoon Missy's virginal attitudes had mentally crossed the Rubicon.) The clumsy motions that distorted their faces could only vaguely be termed kissing, but the horror and delight of the great moment they were about to share made them shake so much that their ineptitude only heightened the passion of the first time.

They were both afraid to speak, afraid that a human sound would shatter the strange trance they had allowed to swirl through and around them. She let herself flow into him. Her face was so hot that it made her kisses and breath seem even cool.

"Yes, Shep! Yes! Now!" she rasped.

He tried to press her with hungry kisses to maintain the mood while the fat bananas his fingers had become fumbled unsuccessfully with, first, the top button, then any button that seemed yielding. The heat from her braless softness seared his hand. Shep was only semi-aware that she was helping him, and he probably only managed two of them on his own. The blouse slipped to the floor someplace.

"I'm so goddamned clumsy," he blurted out in a loud whisper, proving it by popping the snap on her slacks as he ran his tormented fingers down behind and inside her waist to clutch that beautifully molded ass. Missy's small but well-proportioned breasts stood upright in the bars of light that filtered through the venetian blinds, and a large red nipple sprang straight forward into his inclined mouth. He sucked the nipple remarkably gently, considering the wild thoughtlessness of the moment, chewed delicately on it, then tried to suck all of the small, bursting tit into his mouth at once.

Missy pulled back, grunting slightly, but didn't break the warm crystal shield, the cocoon of heat that enmeshed them. Her slender fingers ran from his hips, down his thighs, and across his hard sinewed buttocks before, with little pin-points of pressure, they stroked his untrained but eager hard-on. She dropped to her knees, unzipped his fly, and crammed it into her mouth. Shep knew that if he permitted more than ten seconds of that in the state he was in, it would be all over. He quickly knelt beside her and kissed her with a fever that all but numbed him totally.

With a sideways motion, he pulled her to him down on the rug. He was frantic to get on with it. They could do the other routine any time. The sandals came off easily and, holding them by the cuffs, he tugged at her slacks. Yanking and twisting them, he accidentally flipped her over into a dog-fashion position. She still held him roughly with one hand while keeping her balance with the other elbow. Shep spun around behind her. His trembling fingers pried inside the narrow, damp panty-bridge, parted the vulva, and with her directionless help, planted himself in the wet, tacky nest. It took three or four thrusts before he went all the way in—but he had done it! He was finally actually fucking a girl! Shep exploded almost instantly with that thought and went limp against her, his weight crushing Missy to the floor.

He stopped. Stopped completely. Missy writhed on for a few moments before realizing that she was laboring alone. She didn't climax, but it felt so wonderful that she didn't know how near a thing it was.

They lay drenched in sweat, the sides of their bodies touching, yet seeming not to as they drifted alone and aloft in their own vivid reveries. They both knew very well that it wasn't a polished piece of work they had just performed, but the magnitude of the step they had taken together dwarfed, eclipsed any silly consideration of expertise.

"Did you come?" Shep asked, exhibiting even on his first venture the heavy weight that machismo places on the male ego.

"I don't know. I don't think so. But it was very very nice."

"What do you mean you don't know! Surely you'd know if you came or not!" It wasn't only masculine pride that asked the question. He honestly and tenderly had wanted Missy to have enjoyed, luxuriated, in what was, up to that instant, the major moment of their lives. He had no idea that girls had different excitement thresholds.

"Well, it wasn't the barn-burner I've had it described to me as, but it was far out. I mean really nice. It didn't hurt the way I'd feared it might. Don't ask questions. Let me think about it and enjoy the new me." Missy laughed lightly and said, "It's almost like your joke. I just don't feel that much differently. And still and all, I do. Anyway, we have the rest of our lives to perfect the *Joy of Sex* part of it."

She reached over and gave his hand a squeeze. A short, barely understood wave of fear lapped, and he decided, for want of something else to decide, to get up and have a beer. He was confused as whether he should pull his trousers up or kick them completely off. The idea of tucking himself back in and perhaps messing his pants palled, and he opted to take them off. Vaguely, he was aware that he should kiss Missy or show some sign of affection. But he just didn't feel like it. In an act of discipline, he leaned over, cupped her cheek with his palm, and kissed her dryly. She tried to understand.

Shep's cold reaction to Missy and the cameo they had just enacted was a combination of *post coitum triste,* a disappointment in the first act of love, and the result of his own empty, nihilistic state of mind. He more than partially blamed Missy for participating with him in something that created an atmosphere that was trivial and un-heroic. Love is impossible when there is a prevailing sense of un-reality. The short moment of ecstasy was already forgotten in his desire for something more epic. His own days of wine and roses seemed to have been curtailed to about fifteen minutes, and he felt cheated and angry at Missy. And at the whole fucking world.

He glanced back at her. The large, green eyes in the round head were filmed over as she stared up at the ceiling; her body appeared to have melted into the floor, framed as it was by the deep, shaggy white rug. She too was having misgivings and second thoughts about the love-making, if their crude interlude could be termed that. Looking at her like a camera, Shep was the center of his universe and she was merely someone in it. He was impervious to her reflections. He was too young to realize that Missy, and everyone else for that matter, were each the center of their world and he was only someone passing through their lives.

Shep removed the rest of his clothes and felt a bit better. A little freer. He stood for a second looking at his long, graceful fingers and then down at his toes. He rubbed his body as if it were a stranger who had just been to a foreign land and had something to tell him.

"Are you cold?" Missy asked, sitting up and hugging her knees. Shep focused on her little pug nose for an instant and felt somewhat more kindly toward her, which is to say, toward himself. He swung down the stairs to the bar. It seemed another world down there. The bedroom was no longer a possible enemy to his plans, now consum-mated. Rather, as new charges of sex energy flowed through it, it occurred to him that the bedroom and doing "it" properly might still save the day.

The cold beer tasted like an unknown nectar, and he savored it for a time before calling Missy to join him downstairs. Her wide hips appeared around the corner of the wrought-iron stairs, giving a not unpleasant comparison of cold metal and soft, warm flesh. Shep was inspired again, and his ungrasped rancor of the previous minutes was almost gone.

"What's wrong, Shep? Did I do something wrong? Wasn't it good for you?" she murmured, pressing her cheek against his chest. Shep kissed her forehead and handed her a can of beer. He turned, put his arm around her shoulders, and led her into the bedroom.

"It isn't you, Missy. I guess I just expected too much and should have known better. Nothing's worth all the trouble. Everything's nonsense, isn't it?"

"Oh, Jesus Christ! Shut up, you egomaniac! You—you don't have to spoil it for me too, do you? You're beautiful, you are. I mean, like you've been walking around with this long, draggy face on you for a couple of weeks now. What's your problem? What's the big weltschmerz?"

Finally, and with great relief, Francis Shepard exploded in a torrent of words as they sat naked side by side on the bed. It seemed the proper place, the best person, and the appropriate time to explain his emptiness and fear. He concluded the story by telling her that she was the only thing of any import left in his life, especially now, and that reinforced her own confidence. A strong calm simultaneously relaxed and energized her. Missy wanted to impart some, perhaps all of the new energy to him, and, although one year younger than he, she quite suddenly absorbed wisdom from a source years above and beyond her present plane of consciousness.

"I'm really sorry, Shep. Sorry that this—I guess you could call it sort of a psychological tragedy—thumped you so soon. Before you could have all the ridiculous fun of being young. It's making you an old man without having had the foundation of foolishness kids lay."

Shep didn't rise to that opening for the obvious dumb pun and simply dropped back on the bed, his hands behind his head. Missy was sure then that he was serious. Not yet cynical enough to laugh at the life-predicament. But serious.

"I read once," she went on—"I don't remember who wrote it, but it went like this—I remember it because it made a big impression on me:

" 'Only very young people want to be happy. What we all want is to be quite sure that there is something which makes it worthwhile to go on living in what seems to us our best way, at our finest intensity.' I guess, or I think I guess, that even if a guy, a person, believes that everything is nuts and worthless and shallow, you have to go on living and make the best of it, and I really mean the *best* of it. I don't agree with you, not completely, that everything is as hopeless or use-less as you do, but I do know that all the hard-core realities are pretty absurd. So what about this? Take two things in your life; not necessarily the same two things for always, not for the rest of your life, but for as long as they themselves mean something to you. Then do them with all your energy and love; do them the *best* you can. That way your life will have more meaning for you even though you know they're not actually important in the long run. Have a sense of humor about them and yourself and the whole big can of worms will get it on for you. Be into something as heavy as you can take it, without being a pain in the ass to other people. Does that make any sense to you?"

He had been grasping each phrase and devouring it. Without being a cop-out, it made the first sense to him that anything had in months. Shep thought, perhaps accurately, that it would probably be about as good advice as he would ever get. And yet, Missy's strength surprised and frightened him strangely. He didn't answer. Rather, he took a deep breath and sank more peacefully into the bed. She glanced at him, and perceiving the aura of new calm on his face, an absence of the tension that had been marring it recently, she leaned back beside him.

"Well"—he finally broke the silence—"I know what one of those things I'm going to give my major energies to is going to be."

He rolled over and tenderly enveloped her. They made love, really love this time, and as well as either of them would ever do again, save for a few of the little refinements which later in life would keep it spicy. As they progressed, he ran his tongue slowly along her thigh, only occasionally flicking it alongside her clitoris and then darting it like a little surprise into the pink hole. She was alternately embarrassed and delighted until the heightened plea-sure, the fierce elation of the act phased out all her hang-ups and she whimpered and gasped and sang out audibles she had never even whispered before.

Missy had spread her legs so wide that she was almost aware of how much they ached. Shep was in another world—aflame with what he was doing, yet still conscious that the superior role he had to have at that period in his life was restored. Missy needed him. She needed him to give her the greatest pleasure she ever had. And, in truth, he was doing just that. She was now shouting, writhing, and gasping in a universe all her own. She held him by the back of his head with both hands and called his name, but the experience was the thing, an entity of its own. As she convulsed in the throes of her orgasm, Shep instinctively chose that moment to glide up horizontally and he wedged himself deep inside her.

Tears of delight and unexpected gratification flowed across her face. Shep encircled her neck tenderly and kissed off the trembling drops. He had been so enrapt that he was unaware of the exact moment he had come. They both went limp, but she wouldn't allow him to withdraw. Her arms held him tightly around the waist.

"I love you, Shep. I love you with my whole being."

"Darling," he whispered in her ear, "I love you too."

And at that moment he actually did.

The dry Australian heat crackled on the anthill dirt of the makeshift tennis court. Young Jesse Fraser could feel it gripping his brown, skinny legs as it crept up into his shorts. Even his father had to acknowledge that the boy hadn't natural ability, but hoped that the barber, who had decided to be his coach, would find some talent somewhere that would improve Jess enough to at least lap the tracks after the tennis gravy train had passed. Lenny Staton cut hair inexpertly and his knowledge of tennis wasn't much in excess of the thirteen-year-old Jesse's, but he was keen enough to give his time and to spring for the racquets and balls.

Unlike the great Evonne Goolagong or Tony Roche, Jesse actually did live in the out-back. The painfully white light limned an unreal quality to the barren, reddish terrain—it rolled out flat, before undulating in hot-brick heat toward the horizon. Jesse's father, Richard Fraser, seemed grossly overdressed in his stationmaster uniform, sweat-drenched black wool. He stood with one hand clawed on the chicken wire that enclosed the court and shook his head negatively at the clumsy movements of his son. His son. His hope to escape. To escape down the one track that lay shimmering in heat waves from parallel to single line away and out of this outré outpost of isolation, rejection, and slow death.

Richard Fraser was a big-mouth impressionist who had conceived of the tennis escape valve through Jesse, by the confection he had put around: the simple lie that he was closely related to the perennial wonder Neale Fraser. Jangdongaby had no television. The radio in the local pub made tennis the number two sport to beer drinking in the town, and it was during one of the epic afternoons of guzzling that the stationmaster brought to full bloom the obvious weed. Neale Fraser was in the process of almost single-handedly winning the Davis Cup for Australia.

"I'm not painting the lily when I say that there's talent waiting to be tapped in the Frasers." (One of the obscure facts that had earned him a dollar from time to time was the accurate quote from *King John* about gilding gold and painting the lily.)

No one in town rose to the bait about the lily any longer, and everyone knew he was one-eighth aborigine. Jesse was twenty-one before he discovered that even the name Fraser had been illegally pressed into service by his grandfather on the run.

It was Grandfather Fraser (or whatever) who made his second-biggest contribution to Jesse by snuffing it. His, in this case, timely death, made it necessary for Fraser *père* and Jesse to finally board the rattling anachronism they had flagged so often and begin Jesse's first and greatest adventure.

The trip took four grueling days, but the freckle-faced boy's great blue eyes shone in wonderment out of his small, homely head. Even the change of landscape was sufficient to electrify him with the pristine magic of a world he didn't know. Gliding from the desert into the buttes, then through monolithic canyons that actually seemed to hum power from their sheerness, Jesse was spiritually transported. But the moment of greatest impact was—the ocean. As the train came around a turn, the vast blue on blue struck him, his nose flattened against the train window. The unbelievable wetness of it all—reaching forever, as far as he was concerned.

"Oh my word, Dad! It's a great wonder, isn't it?"

"That it is, my boy. Fair bleedin' dinkum it is." Richard Fraser hadn't seen it himself for sixteen years, and the sense experience made inroads into even his dulled mind. Jesse memorized it, cradling it forever in his remembrance.

"Can I get into it, Dad?"

"Nah! The sharks will eat you. One funeral is enough at a time."

"I read in the paper they had nets up to keep them out," Jesse said with deep disappointment. "I could just run in and out again. Couldn't get me that fast, Dad. Could I?"

"Nah."

They missed the funeral by a day and a half, which was perfectly all right with Fraser—as long as he didn't miss the reading of the will. His old man had bragged by implication, dropped cryptic allusions to great swag, money, or something of great value stashed away. Now was the time, now that he was dead, to "put up or shut up." The obvious contradiction inherent in the thought didn't bother Richard Fraser one bit.

Jesse was stiff-necked from his wide-eyed, open-mouthed gawking at the tall Sydney buildings. He was a gentle, well-meaning boy, and even though his neck pained him in one position he preferred to sit still rather than roll his head around and disturb the august proceedings. They were gathered in a dark little legal office—cavelike in its mustiness, contrasting to the cheerfulness of the robust afternoon sun.

Fraser's half brother, known simply as Mick, was fawning all over the probate lawyer. "Can't do 'im any good," Fraser whispered to Jesse. "It's all down there on a bit of paper."

Mick showed his aborigine strain far more pronouncedly than did Richard Fraser, who had always maintained that Mick's mother had been an abo. There was a world of "brothers" running around the South Seas—Mick had simply struck a tender chord with the old man at the proper moment, wherever, and been taken along as a mascot. Richard Fraser didn't think Mick should get anything at all. "I mean, what the hell is he anyway? I was the *real* son."

The only other person in the room was a distant cousin, an old spinster who kept a rooming house in the poor section of Sydney. Fraser and Jesse were staying there, for nothing, naturally, for their week in the city. Her still young-looking, well-kept tawny hair framed a kind, Fraser-homely face. She was very obviously family.

Dark as Mick was, there wasn't any doubt about him either. The stamp of the old man was smack on his homely face—and the ears. The lawyer looked up from the will to contemplate those eight ears sticking out into the room. They were to plague Jesse for years, before he got his growth, a sharp left hook, and eventually, fame. His head grew into them and they were far less ludicrous in his manhood. But during the growing years, the ears were constantly there, shaming and paining.

From a little old safe, circa 1900, the lawyer produced a large black velvet bag. With a flourish of dramatic importance, he emptied the contents onto his desk. Pearls of all sizes and descriptions tumbled merrily over one another. Giant black pearls. Good quality. Some of them very valuable. Fraser salivated freely, Mick gripped his chair in disbelief, and Doobey, the cousin, stood up and leaned over them, charmed and greedy in her little-old-lady way.

The big shock was to pole-ax Fraser with the reading itself. An old schooner, moored at the far wharf, was Mick's outright. Municipal bonds and a few gold shares totaling twenty-five hundred Australian dollars were divided three ways. But the pearls—the pearls were to be sold, three-quarters going to Doobey and the other quarter divided between the two brothers. That she had nursed the old man through illnesses, drunkenness, and maulings did nothing to dilute the outrage of Fraser. He actually bellowed and grabbed a handful of the pearls. Mick was pacified by the thought of the ship. With his share of the money he'd "outfit her and piss off, carry mail and things among the islands." He knew the ship well. Grew up on her. It was ratty and wretched but almost seaworthy, and she could make time. He knew all too well the dark beginnings of the bag of pearls and was anxious that they be converted into money as soon as possible. Mick knew that all three of the original "owners" of the pearls were dead, but he would still worry until they were cash in his pocket. There could have been a fourth involved in that ugly affair in Fiji. It all had happened so quickly—so long ago—that Mick couldn't be sure.

Mick looked over at the glistening, distorted face of his half brother and howled with laughter. The lawyer, with a soft, patronizing voice, calmed Fraser down and put all the pearls back in the bag, then deftly into the safe.

Fraser had reluctantly released his sweaty bunch of baubles into the bag himself, and the memory of all those black eyes smiling up at him would haunt his dreams to the last. They couldn't possibly have irritated the oysters as much as they were bugging him. "I won't think of them anymore," he thought. Nor would he look at any of the others; still upset but not ashamed, and sensing that he somehow ought to be. He just couldn't remember how to look ashamed.

On the way down the stairs, Fraser resumed his bluster. "Well, that seems to be that. I say, let's go home and have a few frosties. Mick, you black shit-ass, how about we go have a looksee at your Queen Mary. What's it called, anyway?"

"*She's* named the *Coral Princess,* you great white hairy ainu!"

"Your own hairy anus, you great, big, dozey bugger. Come on, then. Lad, do you want to come with us or go with your Auntie Doobey?" (Suddenly she was an aunt.)

As much as Jess wanted to, ached to go to the water and see the boat, he knew all too well that the two of them would finish up shit-faced drunk and fighting, either with each other or with anyone else who cared to mix it with a brawling sailor and ex–sheep de-nutter (he used to bite them off). So Jesse reluctantly boarded a bus with Doobey, while the "boys," ostentatious in their new wealth, roared off—in a taxi, no less.

Beautiful Sydney, from the bus window, was a Disneyland to Jess, and he passed the entire trip in stone-awed silence. When they alighted from the bus, Jesse registered that almost all the houses on their street needed paint. Doobey hurried into the house first. A new boarder, and Jesse's destiny, was waiting in the salon of the turn-of-the-century architectural monstrosity, which, like Topsy, just grew and grew.

Eduarde Vector was a Czechoslovakian-Jewish immigrant (defector). He was assured of a job as a chartered accountant, and as a former Davis Cup star looked forward to some leisurely tennis. Perhaps a few lessons on the side. Sydney was expensive, and he knew he had to save a good deal of money before dreaming of moving to a decent flat. Eduarde Vector's heavy Slavic manner was flavored with enough Jewish wit to intrigue and warm the curiosity and heart of young Jesse. Jesse fell in love with him the instant he saw the five Dunlop racquets balanced on top of the ancient steamer trunk and paraphernalia. Vector saw him staring at the racquets.

"You play tennis, boy?"

"Yes, I'm learning. I like it. It's my favorite thing."

"Yes, *sir,* you mean. Vell, I haf been twenty days on a freighter and need to exercise this constipated body. After supper, ve vill find a court. I care not how vell you play. So don't be nervous playing with the great Vector."

Jesse had never heard of the "great Vector," but he was thrilled to be invited anyway. He paced around the big house until dinner, and itched, squirmed, and rocked in his chair in front of a barely touched meal.

"You might as well go get changed," Doobey said, "I'll put your food away until later. With the prices these days, we can't afford to waste much. When I see that pearl money, that's when I'll believe that pearl money. Right! Off you go!"

Doobey tousled his hair as she cleared away his plate with the other hand. He didn't need a second start. Skinny legs blurred up the stairs, and he was gone for the evening.

When Eduarde Vector and Jesse returned from the park, Fraser was sitting there, more than half drunk.

"The bloody thing will go down like an anvil first time out," he blurted at them, apropos of a daydream he'd been entertaining. "That's to say, if he ever gets out of gaol. Shit-ass Mick stuck one on the harbor agent about back taxes or some such . . ." He lost interest in it. "You've been having a knock with my lad, eh? Good little player. Worlds of Fraser talent in there somewhere."

"You are, then, Mr. Fraser?" Vector stiffly proffered his hand. Whether out of impudence or laziness, Fraser ignored it. Vector immediately made up his mind. He would always hate Richard Fraser.

Snapping his hand back to his side, Eduarde Vector said, looking straight into the boy's face, "It is a miracle that he can hit the ball at all, even the way he does—rotten—with that grip to shovel horseshit. And the stroke! No stroke! If he can hit the ball with that stroke, I can make him a great player. If he works—and you pay."

"Pay! You must be joking! Who do you think you are? I wouldn't even pay Rod Laver—and who the shit do you think you are?" Fraser put his feet up on the small mahogany antique table and roared at the bespectacled, fifty-year-old man beside Jesse.

"I am, you great boor, Eduarde Vector!" This was said with a small rumble in the voice. "Vector" boomed out like a solid mathematical principle.

"Christ!" Fraser half laughed. "I thought you were dead. You look fairly well for a dead man. Whatcha do, go back to ice hockey? Or have you been in prison?"

"If it is some of your business, I haf been in India, coaching and working at my profession, you rude, uneducated pig!"

Even drunk, Fraser was quick. He was on his feet in an instant. Vector took his racquet back sideways with the service grip.

"Not in my house!" Doobey shrieked. "You'll not be crashing about in this house! Now stop this, both of you—at your age!"

Jesse had run up the stairs but was peering between the balustrades on the landing. Fraser dropped his hands and whirled away into the kitchen. Even after the door had closed, Vector remained at the ready, his racquet now trembling. He was only of medium stature, but compact and powerful; still toughened from the years of bone-crunching Olympic hockey.

"I'm sorry, madame. I knew he was tipsy. I don't know what about him made me so angry."

"He'll only be here a few more days," Doobey said with a note of obvious relief in her voice. She picked up some of Eduarde Vector's things and started up the stairs.

The allusion to the few more days brought tears to Jesse's eyes as he perched on his landing. He had just learned in one hour more about tennis than in three years. It was such a pleasure smacking a ball when the racquet was held properly. "And running right and everything."

"What's the matter with you, Jesse?" Doobey bent down over the boy, whose body was shuddering with his sorrow.

"I don't want to go back to the sheep station, Auntie Doobey. Can't I stay here with you a while—with you and Mr. Vector? I want to play tennis—on a proper court. Here," his sniffles turned into a real sob, "in this place. Couldn't I?"

"Well, I really don't know, Jesse. I hadn't thought about it. You *are* good company," she started to think seriously of the possibility, and Jess could hear it in her voice. He started to smile.

"Don't jump to any conclusions, now. I haven't said anything. With the new boarder and those pearls, if anything comes of that—I guess—I suppose I could afford both a mortgage *and* you." She was now sitting on the landing with him, both of them looking down on the room as though it were the world and they had some power over it.

"Just maybe—*maybe*, mind you . . . I'll talk to your father tomorrow when he's sober. Now off to bed with you. Good night, Mr. Vector! I'll have your trunk brought up in the morning."

Vector was rummaging around in a small bag. "Yah, lady. Good night." He waved, turning slightly.

The breakfast table was laden with simple staple food. Eduarde Vector and Fraser sat in heavy silence at opposite ends, with an equally quiet Jesse (afraid to spoil his chances by speaking) and the only other boarder in between them. The fourth man at the table, easily eighty, had long since ceased paying any rent. He was so slight and spent that his tiny body made the long table appear enormous. He talked mainly to himself, but very loudly, proving his existence. Doobey came in with a great platter that washed the room with the delicious aroma of bacon. She placed two eggs, sunny side up, cooked in the shape of a skull, with two strips of bacon crossed under them in front of Fraser. His hangover turned him momentarily green. Doobey started several times to bring up the subject of Jesse, but failed each time. Jesse was silently helping her form the words with all his might and felt that when she failed, he did too.

"Richard," she said, opting for nonchalance, "what do you think of Jesse going to school here in Sydney? Living with me?" The words out, she waxed stronger. "He could get a good education here and he's got no mother out-back anyhow."

"He needs a man's strong hand. A good dad! I don't think much of it."

"The lad could stay with you during holidays and summers and such" (knowing full well Fraser would never go for the price of the train ticket—even half fare). "You could give him all the discipline he needed then. Or you could come here to visit with your pearl money. See how his tennis had improved."

"That man will have nothing to do with my son! I won't have him near the lad. I'll bite his balls off!" he thundered, thumping the table.

The old man's uppers fell into his oatmeal and he farted loudly.

The teetering artificial flowers feel over sideways. Vector washed his toast down with a gulp of coffee and abruptly left the table.

"Then it's settled," Doobey pronounced with finality, seizing on the psychological imbalance of the moment.

"Uh—right." Fraser said this softly, out of breath. "If you want to, boy."

Jesse was careful not to show too much enthusiam. He didn't want Fraser to realize how overjoyed he'd be to part company with him and see something besides the odd dingo, the silent abos, and the three scraggly trees.

Over a cup of black coffee laced with rum, it was decided. Fraser was delighted that he had thought of it. Of course, he'd miss the little bastard—yet, here was a chance to really improve Jesse's tennis, and in six or seven years cash in on some real money. Fraser felt a little sorry for him, though; having to stay in Sydney. "I myself could never stick it."

Staring down at a chip in the plain white china cup of milk, Jesse couldn't suppress a jaw-breaking smile. Two days later Fraser was gone. Six days later there were thousands of miles between them.

Vector knew the fragility of the young boy's stamina and graduated Jesse's conditioning. Short wind sprints; a few knee jumps, increased each week by two; three-pound dumbbells to build up his skinny arms, and a squash ball to squeeze to strengthen his grip. But inside he was ecstatic. He knew what he had for raw material. Even after he had moved to a flat on the other side of Sydney a year later, he continued to travel clear across town after an exhausting day to coach Jesse and to jog with him.

"Wisdom is never diluted by repetition." Eduarde Vector repeated tennis dogma to the boy constantly. "Ven you valk on the court, forget the rest of the world; leave your worries and joys, *and* joys, behind you. Have fun playing but play to vin. Losing stinks! Don't get tight when you play. Relax and concentrate. That means prepare and watch the ball. Some idiots concentrate on concentrating. They are in a trance. When the hell is that, heh? When you play, have a plan Don't just knock the ball around higgly-piggly. Hit it for a reason. Strategy, kid, that's the thing. Ven two players of equal ability play, the smarter one vins. Never mind who is the other person. Play the ball and the veaknesses. Run for everything, even hopeless. You get one back, it's worth three points. How many times did you skip with the rope yesterday? I saw you trying a two-handed forehand against the vall yesterday, like old Jimmy Connors. Connors was a genius. You, you are not a genius. Get the racquet back immediately and prepare. Yah, that's good. Hit out in front of you. Let the racquet do the vork. Stay down, Jesse, stay down. Let the racquet bring your veight through. A little thing like you needs his veight into the ball but not *before* the racquet. *Out in front!*"

They hit thousands of balls in the first two years, and the ritual, the litany, the Greek chorus would follow Jesse to bed and into his dreams: "Bend your knees, kid—step into it, kid—let your hand go where you vant the ball to go, kid—ven you're in trouble on a shot, get it back anyvay you can and get yourself back to the middle—follow the line of the shot ven you go in, kid. Bounce the ball three times for me, Jesse; not two, not four—for me, *three* times before you serve. Then you concentrate and get your weight forward."

For every hundred Jesse hit during the day, he hit another hundred at night in his sleep. Sometimes he'd simply dream of millions of aspirins.

At the end of a practice session, they always jogged for a mile before going home. Vector would give him crits on the evening's performance.

"Vatch the ball on your racquet a little longer. Don't look where it is going. You know the court. Vith your eyes closed you can point to everyvere on the court, so vy you look? Your backhand is gorgeous. Don't do nothing different. Leaf it alone or I kill you, you hear? I vill *kill* you!"

After tournament matches, Eduarde Vector only commented when specifically asked by Jesse. He knew that Jesse knew what had gone right or wrong. Almost always. Win or lose, he rarely said more than, "All right, kid, let's go home."

Jesse met a handsome, chunky (almost fat) boy, first at a fifteen-and-under regional and later at "White City," the vast tennis complex where hopeful talent received nourishment. Al Wick had the deadpan expression of a born heller. His well-made features were maintained in a disguised poker face that would suddenly burst into a handsome, very winning smile at the thought of some mischief to get into. Eduarde Vector all too often had to caution Jesse to "choose a better companion." But Jesse and Wick were healthy and psychologically wholesome for each other. Jesse's drab, wistful personality took on a bit of zest and humor. He emerged from his walled psyche to venture into a painful world with more confidence. Wicko, as Jesse called him, found in Jesse a constraining and sober influence—one that often kept him out of big trouble. Although Wick's mouth was already rotten, Jesse wasn't corrupted by it. One afternoon, years before in the out-back, Fraser had slugged him so hard alongside his head for a curse word that Jesse had literally flown across the room and up against a flour barrel. His head had rung for days, but he had been conditioned. Jesse toned Al Wick down, while inheriting only a few soft off-color expressions, mainly in order to be accepted.

Vector shrugged his shoulders when they decided to be doubles partners, but he saw a good team in them after a month together. Wick was left-handed and tenacious, and they had a good "understanding," covering each other well. Vector coached them both. Wicko was good. Very good.

"Remember, once and for always, 'Vinners go up the middle, losers go down the line!' Repeat that!" They would. "Hit down there from time to time, to keep poacher honest. Not on big points. If you make it, you are a hero—if you miss, you are a bum. The net is six inches, half a foot, higher down the line. Most of the time you vill be a bum. Right? 'Vinners up the middle in doubles; losers go down the line!'

"Al, you try to chip it to my feet as I come in on serve. I vill haf to hit up. Then Jesse, you cross ofer and volley it to hell!"

They tried it eight or nine times—successful only once. Wick had difficulty chipping correctly and only hit two, one of which Jesse dumped into the net.

"Try this, then, Al. Return with topspin up the center. Then follow your own shot in and volley my volley."

Vector served wide to Wick's forehand. It worked perfectly the first time. A looping return that dipped at Vector's feet, causing him to hit up, was crunched away by a backhand volley by the incoming Wick. Then they ran drills, going to the backcourt together on lobs and taking the net together on a ground stroke. They practiced a poach routine over and over. One crossing over from either direction; the other covering, then switching the format around. They were becoming very sharp.

Eduarde Vector had them play a game in which one would lob and the other smash, up to twenty points. Then reverse it for the next game. This not only gave them both the incentive to practice the lob and overhead, it gave Vector a chance to sit down.

One Saturday morning Al Wick and Jesse were practicing alone. During the first set, Wick badly turned his ankle. Jesse went home and found Eduarde Vector and Auntie Doobey in bed together. He watched for a while, confused by the whole thing, especially the cast of characters. He slipped downstairs for some milk and chocolate chip cookies he felt entitled to. Vestor had seen him in the mirror but was amused to have a spectator. Doobey stopped stock still when she heard the door close.

Jesse got a trip to the zoo, an afternoon of the surf throwing him on the beach, and a lecture on sex from, first, Eduarde Vector, then from timid Doobey.

Vector was once again (after a private moment of hesitant embarrassment on the stairs) proud, straight, and mid-European. He turned Jesse by the shoulder and propelled him toward the backyard.

"Gott!" His jaw snapped shut. "That was not for a boy's eyes, especially—for the first time—two wrinkled old tings like us. Vell, boy, anyhow—ven you are older—I tink you will understand better vat I mean to tell you. I tink you will feel, eh . . . Kid, ve all have needs, certain needs, and . . ."

It was sticky going, and Jesse stood riveted in red-eared chagrin, his eyes intently focused on the bird bath. "Actually, ve didn't even made a good ting of it. You scared your Aunt Doobey out of her wits, and I—vell, as my vife used to say, you can't make a good soufflé rise twice."

Vector pulled his shoulders Prussian-tight and terminated the lecture with, "But vat the hell! It is not my business to educate you in those areas. I leave it to your aunt. That should be a goot one, heh?" Vector chuckled at the idea and released Jesse from his little purgatory.

Doobey's lecture was even more painful. She started out with the usual routine, but became hopelessly entangled in a diatribe about the world's injustice to old people when it came to sex. ". . . it is a joke . . " ". . . something vulgar and disgusting." Through this, Jesse retired into his soundless purgatory, thinking, "They could at least lock the door." Besides, he already knew from Wick about sex and babies; how grown-ups took off all their clothes and jumped very hard on each other and after a while the lady grew a baby in one or both of her breasts. Triplets blew his mind. In any case, Jesse preferred Wick's version. The rest was all too much for the virgin brain of a huge child. Jesse much preferred Al Wick's "lecture" on the general subject of sex.

The first visit to Al Wick's home was more than mild shock. Sensitive, prematurely greying, and classically handsome, Wick's father was an eminent judge, renowned for his clearness of decisions and tempered justice. He was soft-spoken and generous of nature. Jesse never saw him angry. At home, Al Wick was an altar boy—all sweetness and bright young boyness. Jesse marveled at the super dissembling Wick pulled off. He was also aware that they had, by some odd twist, exchanged roles in life, with Wicko playing the out-back part. Jess *was* a judge's son in everything but fact. Not that Judge Wick was fooled by his son's bullshit. He had caught Wicko's act dozens of times unobserved. The truth was, he relived his youth (the wild youth he'd never known) vicariously through the boy. The judge had been a serious, thoughtful, bookish young man, not totally accepted by his hard-riding peers. Wicko was liked for his abrasiveness and open manner in the true Australian spirit. Wicko took Jesse home with him infrequently. He couldn't stand the strain of acting for long periods and hated the twinkle in Jesse's eyes when he grooved into his angelic self.

Unfortunately, the occasions of competition between them created anxieties. They were pitted against each other in grueling, bulldog ballbusters. Jesse led in friendlies by only eight matches out of the many dozens they had played. But in competition he was five and nothing. His tactics were sounder, and he had the better return of serve when he focused on it, which he did in tournament circumstances. Their sixth tournament encounter was in the finals of the sixteen-and-under championships. The previous day, Wicko had had a hard, three-set battle with Freddie Moore, but he was nonetheless very fit and ready for Jesse in the finals.

Wick started right in attacking Jesse's forehand, knowing that the topspin backhand Jesse walloped was both devastating and percentage. Wick's left-handed forehand had reverse spin and came down the line instead of cross-court to a right-hander. Wick also knew full well that Jesse was used to playing him—but it was his only chance.

Jesse had long ago figured out Wick's service and, if Wick charged the net, Jesse would put a lob up the easiest place for a right-hander, which also happened to be Wick's backhand. When Wick held up behind the service line in anticipation of the lob, so that he could move around behind it and smash, Jesse ripped off a cross-court, catching Wick flaying and diving and wasting energy. Rarely, otherwise, would Jesse go for an outright winner. Rather, he kept medium pace and played the ball to go moderately deep. He had more patience than Wicko, who too often tried to win the point with one shot. Wick had his moments, though. Jesse hit a short-angled shot to Wick's forehand, then a deep shot to the backhand. Wick was tiring. Instead of going back to the forehand, Jess tried to wrong-foot him with a short cross-court to the backhand, for which he scrambled, lunged, and made a spectacular get, but Jesse put up a lob over his backhand again. Running flat out, back and to his left, Wick changed direction at the last moment, spun right and hit a crowd-pleasing backhand cross-court winner going away from the court.

Jesse stood flat-footed, dropped his racquet, and joined the crowd in long, loud applause. It wasn't long enough, though—Wick had had it. At that stage, in the ninety-degree heat, he probably would have been wiser to drop the point and save his energy. Jesse took the next three games (Wick had to serve after that draining point) and the match, 6-4, 6-3.

"Well played. Bad luck, Wicko."

Wick's handshake was genuine, but the smile was forced.

In their doubles final, a half hour later, the first set was played in total silence between them. Al Wick overcame his fatigue and disappointment, dredged up some energy from somewhere, and together they overwhelmed a first-rate pair that included the doubles specialist Freddie Moore.

They were the sixteen-and-under doubles champions of Australia. Jesse's second championship of the day.

Richard Fraser only learned the good news five days later. He had long ago lost his inheritance in a dicey gold-prospecting venture, and was now "vamping till ready" to cash in on Jesse.

Richard would not have long to wait.

7

It was a little after nine on a bird-chirper of a morning—a day in the "shamateur" years of Randy Mariano. The tight, sharp little rap on the door bore him straight up from the deep regions. For several moments he blinked out at the mother-of-pearl cloud-ripples before fully realizing where he was. A tournament hospitality home. During the hectic, cynical years before open tennis, when 'shamateurism" (amateurs receiving money under the table in tournaments from which professionals were barred) was modus operandi, besides the money, *in cash,* that changed hands surreptitiously, "hospitality" was included in the deal. Good, kind tennis fanatics gave room and board to players for the duration of the tournament, and sometimes longer. Normally the players were polite and well-behaved. But there were also the drunks, the icebox raiders, the wife and daughter seducers, and on occasion, the thieves. Rarely, if you were really unfortunate, there were the Mariano types.

He had arrived several days early, ostensibly to acquaint himself with the surface, but really to rest and freeload. He had had no intention of playing more than one round, since he had entered another tournament for the same time, planning to use whichever one he drew a bye for as the second and real effort. So he had lost in an "upset" to a local player, causing great joy to the city and a lot of gloom to the tournament directors. No qualm coursed through his thoughts as he picked up his money in a book—a gift from the organizers.

Randy stretched his arms and pushed his legs to the limit of the bed, considering his toes in the distance. The knock grew more insistent. Not even in the first semiconscious instants on awakening had he any doubts about who was knocking. For the full five days of his stay in this tasteful, comfortable, upper-class home, Randy had promised his host, Herbert Fruehoff, to play with him on the well-kept private court adjoining the formal gardens. Although Herb Fruehoff was standing at the portals of old age and had a clubfoot, his lifelong enthusiasm for tennis was not only undiminished but honed to a fine, glittering edge. He had long ago ceased to actively manage his large brewery and now spent his time in more worthwhile pursuits like improving his backhand and twist serve.

Every night at the dinner table, while glowing from the heady fine wines, Randy promised the healthy red face, topped with a clump of white hair, to play with him the next day. Something always came up. He never even broached the subject with the Fruehoff boys; picking up a racquet with them was clearly beyond the pale. Eventually their better instincts determined that they ignore him completely, and they were secretly delighted when he lost.

Mrs. Fruehoff knew nothing about the game, and her good nature prevailed over any tensions that might otherwise have become manifest. Herb Fruehoff was undaunted. At eight-thirty, he had hopped from his bed and dressed in his clean, pressed tennis clothes and special shoes. He had his racquet newly strung for the occasion. Mr. Mariano's plane didn't leave until ten past twelve, so there was plenty of time for two sets and a leisurely breakfast of pancakes, ham, and an omelette afterwards. He couldn't possibly know that Randy had a difficult second-round match in another tournament in another city at three-thirty that afternoon.

"How goddamned inconsiderate," Randy thought as he yanked

his tennis gear on. "These rich farts give you a bed and some of the food they have too much of anyway, and they think they own you. What a downer! Not even a pretty girl but an old dude with a gimpy leg. Well! I'll make this short and sweet. He'll think twice next time before he bugs anyone else." Randy swung the door open to Fruehoff's big, happy smile.

"I hope I didn't wake you too early," Herb importuned, placing a fatherly arm around Randy's shoulder as they walked down the winding staircase.

"No, I've been awake for hours. Are we going to eat first or get right to it?"

"I thought we might hit a few balls and perhaps play a set or two and then have a good breakfast by ourselves in the sun room. You know, relax and read the papers."

"I bet they aren't too kind to me this morning, your sports pages."

"I haven't read them fully. Just glanced at the headlines. But no! I wouldn't say they mistreated you so much as they praised our Willy. I think he's going to take an awful shellacking from Kerwin today, though. Too bad you were so off. That would have made an interesting match between you two. We're going to miss you in the finals, too. Not only for the gate money, I'd like you to believe, but this town is starved for good tennis, and we were rather looking forward to good stuff from you. Never mind. Here we are."

They rounded the hedges and emerged from the leafy green canopy to a milk-blue sky. The clouds had dissipated to produce a ball of molten metal that promised to burn the webs from the day.

The ensuing horror was a madman's dream. Even in the warm-up, Mariano ran Herb from side to side with gentle chip shots, ensuring that he could just, but only just, get to them, not hitting anything that might discourage him from running, or rather, hobbling quickly. Before they even began the set, Herb Fruehoff was soaked with perspiration and purple in the face, though smiling bravely. "Well," he gasped, "shall we try a set? Just play your own game at half speed and I'll see if I can learn something."

His white hair was matted damp-flat on his head as he swung into his much practiced service.

"I'll run this old bastard till he drops," Randy said to himself. And that is exactly what he did. With relentless, accurate shots, he ruthlessly nailed the corners with three-quarter pace, never putting the ball away or coming to net to volley. In the second game of the second set, Herb Fruehoff clutched his chest and crumpled to the ground, writhing in the green grit of the court. His moans were muted, trapped between his lips and the ground.

"Oh, for Christ's sake!" Randy thought at the prospect of a huge inconvenience. He walked crisply around the net to Fruehoff's side.

"Hang in there, old fellow. I'm going for help." Herb Fruehoff's tan had gone ashen.

With long strides, Mariano was in the house in seconds. His terse statement about Fruehoff's attack was so sharp and unemotional that it almost induced a seizure in Mrs. Fruehoff. The oldest son jumped up from the table and embraced her tightly for a moment before dashing outside to his father.

While the family was consumed and confused by the frenzy of panic, and doctors and sorrow, Randy Mariano slipped away in a taxi, heading for the airport and his afternoon match.

8

The spring of the fourteenth year in the life of Horace Gray was one of the warmest and most verdantly resplendent in Devon in many an old man's memory. The multihued buddings on the moors swirled into the riverine greens like rose-written Arabic on fuchsia parchment. Horace lived in an alluvial belt just behind the moors and before the sea. Everything in the life of Horace was a sense adventure. He was a beautiful boy who found smells and touch and watching things a total experience. When he pushed his mare at full gallop along the hard sand of the beaches, the salt and wind whipped the hair back from his forehead and temples. When he tried to guide the horse into the breakers, it nearly always managed to dump him into the surf.

The small harbor of Donquay, some five miles from the Gray farm, was a miniature Cape Cod town in appearance—a tiny Provincetown. Its human flavor combined the salty, crusty, provincial humor of the sea people with the laconic, inward passions of the moor folk. Television had made few inroads into the local life-style; it was viewed as a circus from afar, and not as a source of instruction for change.

Consequently, all of the adjectives about Hardy's parochial "Wessex" pertained: dour, passionate, ironic, private.

Horace was discovered swinging a hay scythe with form and abandon. The tennis coach (physical education teacher at the local private girls' school) espied him on a field trip. She was twenty-two and beyond a *coup de foudre*, she thought, but when she saw that sweat-glistening, half-naked young man with the ice-blue eyes, she was struck silly.

"Hello, are we trespassing?" Her round, boringly pretty face spoke with a catch in the throat. Her cream-colored tits were the most titillating he had ever seen.

"No, m'am," he said tersely, addressing her breasts directly.

"My girls—we're from Lady Arder's—my girls and I are on a hike and we've lost the common path."

"Back yonder it is," Horace said down into the canyon of her blouse. Her large, heavy legs were rooted to the knoll. Horace never saw them. Horace didn't even know what color her eyes were.

"You swing that scythe like a tennis player. Do you play?"

"Tennis?" He laughed out loud in his deep new voice at that odd notion.

"You'd be ever so good, you know. Really you would. If you come to the school tennis court next Monday, I'll show you how."

"If," he thought "I come by the tennis courts I'll show *you* how!" He stared at her bolt-rigid-nipples, swallowed dryly, and nodded.

"Good, then I'll look for you."

"Might be," he answered, swinging a hot scythe through the grass. He then noticed the other little possibilities moistening themselves over him.

"Monday—why not."

Giggling and peeking over their shoulder, the girls moved off along the footpath toward the moors. Horace attacked the hay with a lusty song, occasionally thinking, "Tennis?" and knocking himself out with blackbird-scaring laughter.

Football and cricket were the big sports of the village. Donquay had never produced a tennis player who was even qualified for County Week. Young Horace donned his white football shorts, wrapped a pullover around his neck in case the wind shifted, and decided to take a horse instead of the van. He knew that a man looks twice as good on a horse and half as good in a car. To be on the safe side, in case he met any friends, he carried his football shoes.

The school lay back in a cluster of trees between the farm and town. The walled side opened onto a village street. The school was small—one classroom building, a combination residence and administration building, and the dormitory.

He trotted gently, easy in the saddle, among the buildings and through the trees until he heard girlish squeals and the popping noise of tennis balls. The court was grass, but damp earth surrounded it. He reined up the horse on the other side of the court where he had glimpsed a lush clump of grass. Although high branches shaded the area, Horace had to jam a rock in each of his football shoes and tie them to the reins as an anchor.

He left the horse to graze on what was the new growth, replacement grass for the court, and strode barefoot to the fence. His mentor had just walked out on the court and, with a whistle which had been concealed between those two gorgeous groodies, blew finish to the day's happy grab-ass—for by no stretch of terminology could the bounding, lunging, or screaming of the hopelessly enthusiastic girls be called tennis. They ran off noisily toward the dorm like long-legged fillies.

"What a lovely surprise!" the older woman said (she was an "older woman" to him). "Really it is. Oh, I'm so glad you came—Crikey! Your horse is eating our court!" She ran off and moved the horse away from the turf.

"You're a devil. You really are," she said when she returned, taking his hand like a fellow conspirator. "Come on then. Shall we take a crack at it?" She led him to the bench where racquets of all sizes, weights, and ages were strewn. Collectors' items most of them, some practically triangular, with wooden grips; others had metal strings gone to rust. Few of the balls had nap on them, but Horace wouldn't know that that wasn't the natural condition of tennis balls for several months. As he handled the strange objects like hot potatoes, she sized him up with sidelong appraisal. He hardly cut the dashing, sexy figure she had remembered, standing there now in what seemed to be his underwear, looking young and oafish. Yet the eyes were divine, and even in his unfamiliarity with the milieu he had a certain *je ne sais quoi*. He picked up two balls and playfully stuck them inside his T-shirt, forming small breasts. She looked at the dormitory—right through him. Even at fourteen, he was shrewd enough to realize that it would only take a few more silly turns like that before she would discover his age and the whole thing would be spoiled. To never be able to mold those marvelous hot melons to his face! Maturity—that was the thing. Slowly, he lifted his shirt and the balls tumbled out, struck together on the ground, and bolted away at right angles.

"My name is Sirena," she informed him. "Don't you have any tennis kit?" He had a good handshake. Strong pectorals.

"Tennis kit? I don't play the bloody game, do I!" he said. It was a declaration, not a question.

"Never mind. Not to worry. I don't think anyone will mind if you play without shoes just this once. Tomorrow you can buy some."

Easy enough for her to say, he thought. Got better ways to spend me allowance.

Expertly, she handled the racquets for weight and balance before selecting one for him.

"I'm Sirena—didn't hear your name."

"Horace" lodged in his throat somewhere near the esophagus as he tried to bring it up deeply. A "Horace" was never going to get anywhere with a "Sirena." He looked off into the distance through the trees. There was a Mark-Nine parked by the administration building.

"Pardon?" she cocked her head slightly.

"Jaguar," he near-growled.

"Oh—what a *super* name! Now then, shall we?"

Thus "Jaguar" was born—with a gallery. The dorm windows were crammed with spectator-voyeurs. Sirena showed him the grips. As she turned him to demonstrate the proper way to take the racquet back, the back of his hand felt the hot perspiration through her shirt. He assumed a ridiculous stance in the middle of the court, and she fed him balls underhand. He whiffed the first one—moans from the windows. But the second went over the fence for six and the applause resounded through the trees. The ball carried beyond the horse and down a gully toward the brook.

"We'll find it later," she said with cheerful magnanimity. But her thought was: the hatchet on the piano. Sirena made the finger-to-lips gesture for silence toward the cheering girls and they disappeared, back a step from the windows.

After ten or twelve balls (they both had racquets now), it struck Sirena that they were having two-and three-shot rallies. She concluded that he surely had played before and was having her on. Nonetheless, he needed worlds of practice and instruction. After a half hour of whacking the ball all over the place, the girls watching from the dorm became bored and were off to other entertainments.

Jaguar was bored too. Walloping the ball all around the lawn was, to Jag, simply a means to an end. Sirena had the same thing in mind, about twice as intensely as he did, and they called it a day simultaneously. It was getting too dark anyway.

"I'll help search for the ball I lost." His excitement swallowed the last few words.

"Yes, let's."

On a flat, mossy incline between two large rocks, Jag struggled his wang up the leg of her shorts, inched her panties aside, and fucked her clumsily. They were both too hot to notice the rough inconvenience of the position, but, after they had come, like wonderful animals, quickly and completely, they were slowly aware of soft, stifled laughter among the gurgles of the brook.

"Damn!" Sirena whispered. "Those nasty little bitches are lurking around down here. Maybe they couldn't see." The steam from their bodies was drawing mosquitos, which swarmed in a pre-attack dance just over their heads. "Do you see the ball, Jag?" she called loudly right into his face. Outright laughter this time. "Shit and damn! I'd better catch them!" With startling strength she rolled him off and out of her, and Indian-fashion, stalked away across the brook.

Jag sat up right into the mosquito caucus. He tried to wave them away with arms and hand swats—then decided to retire with grace. Hearing some angry conversation from the other side of the stream, he stood confused for a moment, then climbed up the knoll to his horse. Halfway up, he luck-kicked the ball out from its ground cover and took it home with him, tossing it in the air. He made up snatches of a new song as he went:

"Down my leg, leg,
That's what legs are for. . . ."

As he sat astride his horse, she called from the edge of the hill, "Tomorrow—second lesson? And surprise?" He waved and nodded. Two girls in uniforms appeared behind Sirena. They smiled and waved too.

He galloped through the trees and out across the fields toward the darkening moors, singing:

"Down my leg and up her leg,
Hey nonny non. . . ."
Hey nonny non. . . ."

Once again Jaguar imbibed the rich scents of night at full tilt, and was intoxicated on pure life, as high as he ever would be on pot. Before going into the house, he rubbed the horse down well and set it loose. The house was incongruous with the rest of the farm complex. The original building had burned down soon after they had moved into it from London. Jag remembered the fire only in the form of an infrequent nightmare.

His half-American mother had been an exchange student at Smith College in the United States for two years. During that fecund segment of her life, she spent the greater part of two summers with her roommate on a beef ranch in southeast Texas and never forgot the simple but sumptuous one-story, L-shaped ranch house. It was an easy matter, with the high-powered American side of her, to sell the idea of the house to her gentle, line-of-least resistance husband, who didn't much care what he slept in as long as she shared his bed and he could get on with the business of the farm. He loved her very much—as much as she adored him. Driving up the winding road to the house, he would sometimes find his face in an involuntary wince as the house came into view. He was certain it looked perfectly all right in Texas—but Devon?

John Gray was Devon-born, on a farm not unlike and not far from this one. At first it seemed that Oxford was an escape from the shackles of the farm and small-town life, but after coming down from University to begin the London business whirl, he found that being a Latin scholar and literary buff didn't quite prepare him for the confinement, petty practices, or inexperience inherent in the life of a business novitiate. He met, fell in love with, and married Miriam, who was very keen on farm life. Thus he escaped back to the freedom of fresh air and priceless, reasonable refuge; the soundless solitude of the empty barn, the universal unity of the predawn sky.

"You're late, Horace. You have school work after dinner, so don't dawdle." John Gray sat benignly at the head of the table, an unlit pipe in his mouth. The shock of sheer white hair that topped his handsome head was testimony to the late arrival of Jag's little sister, Amelia. She ate in a silent world of her own creation. Even while eating, she was led off on incredible journeys by her imagination. At first, when she had been much younger, her parents had feared that she was deaf because it took her so long to return from her trip when spoken to. Even now they still spoke to her with inordinate emphasis.

"If you've finished, Amelia, you may be excused." Her glasses seemed most of her face, and she was awkward. She wouldn't always be. Amelia had a waiting swan.

Jaguar was awed by his father, by his strength and knowledge. This awe was a strong element in the inarticulateness of Jag's later life. They only spoke of things pertaining to the farm or school.

"How's your schoolwork going, Horace?"

"Not too well, Dad."

"Don't fret, boy," he said with some irony. "There is a potential thunderbolt in the clouds of destiny, waiting for the right moment —the time of need."

"Mine must be saving up for a big one," Jag said ruefully.

"Don't worry, son. Get the smell of young dollies out of your nostrils and your levels of concentration will soar. You didn't do your chores this afternoon."

"I'll do them in the early morning, Dad." Then, with a hesitation: "I learned tennis today."

"Oh, yes? Well then. That *is* remarkable." Although there was a definite tint of sarcasm in the statement, there was enough pleasure gleaming through to cause Jaguar to continue from strength.

"I learned from the school mistress at the 'Lady's.' Said I could do it if I really enlisted myself."

"Applied, you mean. Fine, lad, I don't play the game, but I've always enjoyed watching it. You couldn't be clumsier at it than you are at football. Says you could be good, eh? Give it a go, lad."

Jaguar smiled inside, thinking with renewed warmth of the bevy of young twitty-twats, the little "perks" of his new dimension. He went at his vegetable soup with orgiastic fervor.

ABOVE-AVERAGE INTELLIGENCE

GOOD COMPREHENSION WHEN ATTENTIVE

BAD ATTITUDE

SHORT ATTENTION SPAN

LIKES HISTORY

Could one of you please arrange to visit me to discuss Horace's work?   —Jane Newberry

This page of Jag's end-of-trimester report burned a hole in his brain as he considered the relative merits of ripping it up, running away from home, and suicide. "Fuck it!—I'll take it as it goes. I'll be fifteen in a month and that's almost sixteen. Then I can become a pirate and rape girls by the boatload."

It was the "Bad Attitude" that was going to cause one of his dad's rare blow-ups. "I'll be good today at tennis. I'll have a good attitude at tennis today." His father was off in a distant pasture searching for a calf. Jaguar put the odious cardboard square in his saddlebag and lit out for his next "lesson."

Sirena's "surprise" was in the form of delicious, sugar-coated blackmail. The two precocious hellions, the source of yesterday's knowing laughter, had threatened to reveal the "Secret of the Brook," as they termed it, if she didn't share Jaguar. He was transported by the whole idea and glanced hungrily and often at their virginal faces and deceptive bodies. They sat on the courtside bench, staining their gym clothes watching, paralyzed with anticipation. They were in fact virgins. They had made an adolescent pact. A smuggled book was studied in depth, devoured. It was only soft porn, but it had brought them both to a decision. This was their moment of truth. Jaguar fulfilled the high qualifications, and they had reached the predetermined time limit.

Their combined nervousness and desire was transmitted to him— he had plenty of trouble concentrating. Sirena was of several minds—

jealous and randy at the same time. Jaguar was trying hard to put them out of his mind for a while and get Sirena to remark on his attitude. The word was an *idée fixe* now. Industrious, natural, well-coordinated . . . these words just didn't cancel out the indictment in the saddlebag.

"How do you like my attitude?" he blurted out, after a long run and complete whiff.

"Splendid," Sirena chuckled at the non-sequitur.

"Splendid what, bloody hell?"

"I like your flaming attitude," she said, with the beginning of anger.

"Oh, do you really?" Jag's relief came over as false modesty and rankled Sirena.

From the bench one of the girls knifed in with a shout: "Are you two going to knock up all day? Play and get it over with."

"Don't be impatient, you silly little moo. He doesn't even know how to keep score yet—we're almost finished."

"I can learn as we go along," Jag chimed in. "With my good attitude and good comprehension when I pay attention."

They played five games, and to her amazement he actually won a game by hustle, speed, and luck.

"You are really exceptional, Jaguar! Very talented indeed. You have a natural gift."

"Will you write that in a note to my father?"

"Yes, of course, but why?"

"Then he'll allow me to come over here for lessons. He thinks tennis is a waste of time."

"Jaguar, love, I'd be very glad to."

"And please mention my good attitude and sunny disposition."

Sirena didn't know whether or not to take him seriously. On second thought, she could see that it meant a great deal to him, and she was quite flattered.

Sirena had the key to the gymnasium–recreation room. The small first-aid room in the rear, little more than a large closet really, had a military bed in it with a lie-low (an air mattress) on it. When it was dark enough, they filed into the dark gym, the girls excited with the mystery of it, the forthcoming escape from the "chains" of their virginity (their phrase), hearts beating wildly over the whole conception of the plot.

Jaguar was more than passingly interested himself, but his excitement was somewhat diluted by confusion about the mathematics of the arrangements. He need not have bothered. The students had devised a master plan long before they had ever met him. Bad luck for Sirena who didn't fit into the scheme of things, but, in childlike innocence devoid of any arrogance, Jaguar whispered to her, in the steady tones of loyalty, "I'll save the best for you, m'am."

The first girl led Jaguar into the pitch-dark room that smelled of gauze, iodine, and a hundred years of skinned knees. She had only the marginal apprehension of a June bug about to be zapped by a duck. He started to pull off his new tennis shorts. She stopped him. "I'll take them off, please? We've planned." The idea added a quarter of an inch to his rod.

"What about what's-her-name?"

"Oh—you have to put them on again."

Jaguar thought that was silly in the extreme, but was committed to do it their way.

"May I take yours off?" This with a sweet deference—quite new to him. There is something about dealing with quality that makes even the hardened hesitate. Jaguar had been drifting into callous shadows lately. The religious aspects of the ceremony dawned on him like a small light in the sacristy of his awareness. A candle was produced from her bag. He hesitated, not knowing what was wanted of him, or rather, how it was wanted.

"If you don't want me," she stammered, misinterpreting his indecision for reluctance, "you don't have to take me. But I promise you, I'll cry a great deal."

They laughed nervously. He took her things off in instants—shorts, pants, and tennis shirt. Her nearly developed body had a radiance in the candlelight. Jaguar thought: These aren't just two mysteries what wants to get screwed—there's more here.

The girl's lower lip trembled as she touched the waistband of his shorts. She seemed on the verge of hysteria. Jag began to consider strange possibilities. Witch covens were a very frequent fact through the countryside, growing more and more popular. "Sweet Mother! I do pray I'm not into one of those!" He crossed his fingers with that thought.

The buttons undone, his shorts fell to the floor, revealing him as an enormous arc. His imagination, his fear had delivered him into a half-mast state of confusion. The virgin busied herself with finishing disrobing him; she had never seen a man's cock before and wasn't sure what an erection should look like. She was pristinely lovely in the glowing candle, and occasional glints of moisture on her inner thigh proved the scene to be more than just a cerebral ceremony to her.

She opened a bottle of baby oil and began shyly to apply it to his body. Her movements became less tentative and more and more finger-flexing as she lost her fear of his strange body. He was shimmering, dappled in the flickering light. She, with ultra-timid motions, applied the oil to his cock, first with manipulations that were over-taut, then slower, and with a tenderness meant for sensitive things. Jag was so excited by it all that he was convinced that he could feel his pores opening and closing. She handed him the bottle. First with hesitant respect, then with more rhythmically plying fingers, he applied the oil. When at last he glided along the white satin of her inner thigh, she moaned out loud. He faltered a second, then slipped a finger inside. She shuddered—her entire body shuddered—and she pulled him to her and buried his lips in her mouth. Her tongue darted in and out like an involuntary muscle. She held him like a crowbar. Slowly, with the silken texture of the mood, Jag pried her fingers off him, moved her back onto the bed, and spread her honeyed lips apart.

"Ouch!" she cried. "Oh, I say! That really *does* hurt. Don't stop. Go slow—yes—Oh, God, that is—ow—oh—more terrible than—oh—oh, yes!"

He exploded, very prematurely, into a million pieces.

The young thing was only a little disappointed, not having known quite what to expect. Yet, there was the profound perception of a void in the plan. She knew now that she shouldn't have done it without love. Her friend Sally may be capable of a pure sexual liaison; it was Sally's plan, after all. But she was now very conscious of her own need for love in the act. She finally allowed him to withdraw, dressed, and thanked him.

"I'm not the wild devil I thought I was, but at least I'm not a boring virgin any more. You see . . . these are our coming-of-age rites." With a kiss on the cheek, she left him in abysmal confusion.

"Daft—that's what they are. Bonkers!" Jaguar put his shorts back on for Sally to remove, wondering if he was up to it. "Lord knows,

I don't need any more oil." When she came in, it was instantly clear to Jag who the leader was, who, perhaps, the witch. Sally was smirking with the Satanic leer of a walking orgasm. Jaguar got goose bumps. "I can always punch her up the throat if she gets witchity," he thought, steeling himself.

She held his head in both hands and kissed him, building up in intensity gradually. Then, still kissing him, she removed his shorts and ran her nails up the inside of his thighs. When he reacted with a quiver, she knelt and sucked him, inexpertly but with enthusiasm. He was rigid again. He was proud. He wished that he'd brought a stopwatch. From her bag she produced baby oil. Jaguar's eyes rolled to the ceiling and he said, "I'm a quart low on the legs." Sally looked up, startled for a moment, then continued, slowly, gently rubbing him. "Lie down," she whispered. He shrugged his shoulders and obeyed. She replaced the oil in the bag and brought out another jar. When she applied the contents to his toes, they seemed to freeze. One at a time she sucked them to life, sending tingles up his leg. The toes were for openers. She sucked everything else as well. Jaguar wondered, in rare moments of lucidity, if she would go after the chair and table legs next. Now her great moment arrived. Carefully, so as not to pinch him anywhere with her knees, Sally mounted him, gently, gradually, allowing herself to settle around him. It was difficult. She mistakenly believed that his prick was normal and not the freak it was. He actually spoiled it for her for years to come. Her quest was begun as of that night.

Although he was delighted from time to time, Jag couldn't help reflecting that the dairy girls were far more fun and less confusing and . . . well . . . he was in charge. Sally was humping wildly on him; painfully tight at first, grunting in pain at first, then she was away.

"Roll over on top of me," she commanded. That done, he felt better about everything. "Talk nasty to me, Jaguar; be disgusting."

"You whore—I'm fucking you, you pig—oh you silly cunt, you! I'm going to rip you apart with my big prick." He was warming to it; in fact, as he seldom swore at all, it was a form of purgative to him, and they sullied themselves in total delight. By sheer coincidence, they came together. He spoiled her the second time of the evening. Something else for her quest.

When her exhausted, breathless stage had passed, she took a tube from her bag and anointed, first her clitoris, then the head of his cock, describing a cross with a circle around it.

Jaguar crossed his fingers and waved his hands back and forth to avoid evil eyes. His own dick, with the red circle drawn around it, looked like an evil eye. Sally kissed his eyes, both ears, and mouth, mumbled something, and walked over to the candle to dress. It was only then, dispassionately, that he realized what a splendid body she had. It was as if it had never been near him at all, glowing with line and form, unspoiled—and in a sense it hadn't.

She clasped her hands in front of her face, bowed to him, and left.

"Great thundering horseballs." His furrowed brow ached from the attempted comprehension. When Sirena entered, saying, "I'm here for the best," in a coy voice, he answered, "I'm only here for the beer," and fell limply back on the couch.

Five months later, Jaguar had fucked his way through a third of the school. The student body bought a huge stuffed Jaguar for a mascot and took it to all the sports events and social functions. Most important for Jag, his tennis had become a progress marvel. Sirena could no longer win a game from him—he had become officially her paid assistant at the courts, and under Sirena's tutelage he won the first two fourteen-and-under tournaments he played. Minor events, yes, but a solid test against boys who had been playing for three and four years.

In the third tournament, in Torquay, he lost, but caught the eye of the regional coach, who shortly after paid a visit to the Gray farm.

Ferdy Archer was one of the three best tennis coaches in Great Britain. He ranked along with names like Mottram, Davidson, Iles, Pearce, Lowe, and Upton in the annals of English tennis. Naturally, Jaguar's father had never heard of him. Ferdy was a little disturbed at this in the beginning of their meeting. It always helped to "go in famous" no matter what the walk of life. To be anonymous somewhat weakens a winning position in salesmanship.

"Mr. Gray," Ferdy said; he started to light his pipe, hesitated, then continued at a nod from Gray: "I don't know how much Jaguar is needed here on the farm or what his school situation is, but—and I'm not being completely unselfish in this myself—last month I received a rather frantic letter from the LTA headquarters in London. They asked me to send any good, young sixteen-and-under prospects up to them straightaway. Now, I would prefer to have him over in my bailiwick for a year at least. Polish him up . . . instill my system of play and court manners, but they want to show some results, Christ . . . I don't know . . . some value for money . . . up there to their directors. Dearth of talent, that sort of thing there. Personally, I don't think he's ready for their factory yet. Still, he *is* the best hopeful I've seen down here since Mike Sangster. In all good conscience, I *have* to recommend him. Well—what do you think of all that?"

"I quite don't know what to think, Mr. . . . ."

"Archer. Like the 'Archers.' "

"Mr. Archer. For instance, what does this 'factory' you mentioned entail? For my own part—and I've worked and lived up there—that word sounds very depersonalizing. Now, I don't need *Jaguar*," he laughed. "That's what he calls himself, is it? I don't need him to work here. My dairy is doing nicely, thank you, and I can afford a hired man."

"Factory was a badly chosen word. What I intended in it was the reality that they work their small, highly selective group very hard indeed."

"His schooling won't lose anything by his changing over. For that matter, neither will his school. The change could only be an improvement. You say he's not ready—yet you say that they need him up there. That doesn't speak very highly for the caliber of British tennis."

"The caliber rises and falls like anywhere else. But we are a small body, the weather is against us, and America skims off the cream with contracts and other inducements and attractions as soon as we groom anyone."

John Gray picked up his pipe but didn't light it. "Rather like shoveling shit against the tide, isn't it?"

"Exactly my sentiments, but shovel we must until we work out a scheme of our own. We've fought them too long and lost. A flexible compromise and an imaginative format long ago would have solved most of our current problems. Only people like Rothman's and Dewars and a few newspapers have saved us so far."

"For a start, let's try this on. I have relations in Richmond. He could stay there and, no doubt, there's a school nearby that might have him. But I want him to work with you through the summer months and go up after that. I don't want the lad destroyed to satisfy some group of letters. All of this is speculative. I have to ask Mrs. Gray first, of course. I may have the last word on the subject, but that, Mr. Archer, I assure you is only a formality. It's her word that counts."

"Quite right, too." He stood up to leave, "It's a great opportunity for the lad. He has such natural talent. Of course, it's a big jump from the juniors to the big time, but he ought to have a run at it."

"We'll talk it out. We both love the boy. He's not the brightest and already shows a proclivity for following his John Thomas through life, but he's sound. He's there when he should be. Thank you for all of your interest."

They walked to the door and shook hands.

"It's my pleasure. I think you'll find we're going about this right. I'll be delighted to have him this summer."

"Goodnight to you, Mr. Archer."

"Ferdy to you. All the best."

Jaguar wasn't overjoyed at first to leave his built-in harem, but the idea of London frothed through his blood. The girls, on second thought, were making heavy demands on him anyway, and outbreaks of jealousy were becoming more and more frequent. Sirena wasn't even speaking to him anymore and referred to him as her Frankenstein. "I created a monster," she often wailed. There was quite an active cult surrounding him, led by Sally the High Priestess. It was even beginning to frighten him. "London! Aye, London—that's the stuff."

With Ferdy Archer in Torquay he worked, put out major effort. At first the eleven-o'clock curfew was maddening, but he managed to have it off with the maids at the Palace Hotel irregularly in irregular places through the day.

At fifteen, Jaguar Gray had achieved most of his height. The farm work had strengthened him, especially his wrists. They were as large or larger than the ordinary man's. With Archer's help, he had power and depth on his volleys and felt at home anywhere on the court. He was quick. Jag had the fleet agility of Tom Okker and the physical capacity to bludgeon the ball that Lew Hoad wielded—deadly and often. His strokes flowed naturally, and Archer didn't have difficulty adding the little bits of finesse, the touches that raised a shot from adequate to near perfection. On a tight volley, hit almost at him, Jag had perfected a quick flamenco, a matador motion, leaping to his right, and backhanding it, or around to his left and hitting it behind his back. The latter was a crowd-pleaser and, for him at least, a high-percentage save-shot as well.

Only his serve was inadequate, and that was a monumental stumbling block. It was wild and hairy, since he persisted in slipping around and out of the proper grip. He had gotten away with serving with almost the forehand grip against hackers, but now Ferdy knew that he would have to force the change if Jag was to entertain any hopes for a tennis future. Nor did Jag throw properly. When he did infrequently throw rocks as a boy, it was always side arm, and bowling the cricket ball was the wrong motion altogether. The pitching movements in baseball, snapping the wrist at the top of the throw was the perfect series of moves for the serve. They didn't play that much baseball in Devon.

However, he was striving, and Ferdy Archer was slowly sowing something at least equally important: the will to win—the dogged insistence on coming out on top. Constantly he repeated the American maxim: "Show me a good loser and I'll show you a loser." It had nothing to do with behavior after a match. Jaguar had some grace in that respect. No, it was the attitude *during* the match that it did matter who won. The "it's-only-a-game" concept lost the English many a game in days gone by, but the new breed was mentally getting it together and showing heavy balls in the crunch. Not that the British weren't good in the crunch; that was their best and most dangerous position. Simply stated, they just hadn't in former times considered a critical point in tennis a crunch. They wanted the results in their favor, yet played like fairies.

Only the few who had their eye on the juggler: Fred Perry, Austin, Taylor, and Cox—Wade, Mortimer, and Jones, to name a few of the few—only they brought home the honors that the British cried out for. Jaguar had it. Like Cliff Richey, he would run for anything and never quit. Jag was not lazy, and since he was good at the game, he liked it and wanted to improve. He could literally taste a win. There was a different flavor in his mouth when his racquet went up into the air after the last point. Something chemical.

The summer had glided by like the kiss of a breeze on a leaf. The time for Jaguar to go up to London arrived in a rush that took them all by surprise. Miriam, his mother, adjusted as bravely as possible to the trauma of losing a child to his own time—and world.

She abhorred driving in traffic. John Gray found it more reasonable to humor her quirk of driving late at night to avoid it. Miriam's eyes teared as she noticed the cold alacrity, the detached enthusiasm with which Jag leapt into the car without a look back. She sighed heavily, trying to remember the moment of her own break, and climbed into the car beside him.

John Gray kissed her through the window, and they said all the unnecessary things that happily long-married people use to cover silence. He was staying behind with two sick cows. He didn't enjoy driving anyway, and London held no pleasant mysteries for him anymore.

Miriam Gray allowed Jag to drive in stretches, over isolated parts of the moors, but she was afraid in populated areas without L-plates and reassumed the wheel. It broke up the six-hour drive.

They rolled through a quiet, ghostly London at five in the morning, not foggy but with enough mist to dampen the spirits of Lord Nelson in a sepulchral Trafalgar Square. From the Strand up to Piccadilly, the streets were all but deserted, but Jaguar's sleepy eyes found excitement in just being there. Store windows glittered along an otherwise uneventful Kensington High Street, hypnotizing Jag with the magic of it all. Miriam's old Mercedes droned along past Edith Road "leading directly to Queen's Club" and across Hammersmith Bridge. The curtain on Act Two of his life had just gone up.

Miriam Gray stayed only two days for shopping, but during that time she had tea with the chief LTA coach at the Queen's Club, a jaunty Yorkshireman named Miles Berryman. She briefed him on all the background material she was convinced he would need to handle "Horace," as she called him.

". . . he's a handful. Keen on tennis and his eye for the girls? Yes, that's pretty keen as well. I can see that you're a man of a great deal of experience with young people from watching your workout session. Yet, wisdom and compassion need clout to be useful. I want —I give you permission to kick Horace up the backside whenever you deem it necessary." She sat back with a rush of breath, her part over, she felt.

"But you see, my dear Mrs. Gray, I've been trying to tell you that

I'm retiring. Didn't anyone down there"—with a wave in the general direction of south—"tell you that Ferdy Archer has been appointed as the new head coach for English?"

She was surprised, flustered, and delighted. Flustered because she didn't want Berryman to be insulted at her delight and tried to hide it.

Berryman conducted her on a tour of the unimpressive complex. Only the marvelous grass courts, green and playable even in their last week, brought a nod of admiration and pleasure.

Before departing from Richmond, Miriam gave her son all the advice and admonitions mothers universally dish up. With a kiss and a solitary tear, she said good-bye forever to his childhood and went out of that segment of his life that was over. It was a lonely and reflective drive back to Devon.

Jaguar's reunion with Ferdy was one of happy relief. Archer went through the obvious routine about showing no favoritism, but—unless there were revelations among the group, which he doubted in view of the national summons to coaches for talent—Jaguar was *It*, and that's where his emphasis was going to be applied. Reverse favoritism it would be. Jaguar was going to have to work his ass off.

Hours against the wall, endless running, drills of threes, boxes —wind-sprints against a stopwatch, in which six balls were placed against the squash court wall and the boys and girls ran from wall to wall, collecting them and putting them into a box. In practice matches, Jaguar lost the first twenty-three: his service was miserable, and so was Jag. Slowly, his spirit was being extinguished. Losing didn't figure in his scheme of things, and although he seemed to take it well on the exterior, Archer knew the boy was being crushed by it and worked extra hours on the serve. Throwing a hundred balls, then serving a hundred balls, coaching, swearing, exasperated until the final breakthrough, when, like a man with cataracts removed, Jag could see—the rhythm was there—the timing was on. Not always, of course, but there to be reached for with increasing frequency.

When Jag won his first set, the confidence flowed slowly and warmly, coursing an inner peace through him. He was a strong boy with a strong will, but another week the way it had been going and he had been quite prepared to pack it in.

Winning became a more common occurrence as the serve pulled

the rest of his game together. Two months later he was steamrolling over everyone, and his presence on the court was a total thing. His manner in winning was outwardly just as amiable as it had been in losing. So Jaguar was liked by his peers and slowly assumed a position of leadership among them. Archer was in raptures but maintained his strict demeanor toward Jaguar. This was the most fragile span. The dangerous binge of early success. Many good ones finished right here, struggling no further; content to be a big noise for a little while.

The year passed quickly, if arduously, for them both. Jaguar beat Billy Sherman with ease to win the sixteen-and-under indoor championship on the wood, and added three more cups in the springtime, playing on clay. He was almost as famous on the junior circuit for the size of his cock. The girls even went to the extreme of bribing some of the young men to spy on him in the shower to see if it was true. They all corroborated the now rampant rumor. His position in British tennis was assured, even if he never won another thing.

Jaguar made no dent in the curricula of his new school either, but he played number one on the tennis team and that gave him a leg up on otherwise lackluster grades. His only "A" was in history; doubly rewarding because his instructor invited him to the staff's breaking-up party at the end of the school year, and Jag managed to get the instructor's wife off in a classroom where they made a little history of their own.

That summer, after a string of wins around England, taking a place on the King's Cup squad, which won in Holland, and a good first-round win over an older, seasoned player in Monte Carlo, Jaguar capped his new fame by very nearly winning Junior Wimbledon, going 12-10 in the third set to an Australian boy named Fraser. It was all the more impressive a showing because Jesse Fraser was, even at sixteen, a veteran with experience and reputation. Jaguar had been playing a little over a year. Somehow, the loss pulled him into a decline and he made little headway for the next year. In fact, he took a summer off from tournament play and returned to Devon to work on the farm and relax. Archer was violently against it and ranted long and loud over the move, "smack in the middle of the season." Jaguar answered that nowadays the year seemed to be one long season. "I miss my folks and I'm going home."

It was salubrious for everyone involved. The other boys received a more equitable share of attention from Archer so that the English team wouldn't be a one-man band; Archer lowered his blood pressure several degrees, no longer clucking and fretting over Jaguar's performance; and Jag got the greatest antidote in the world for homesickness—going there. He was glad to see his mother and father, and they him, but after a week it was obvious to all of them that a great change had been effected in their lives. He had made the quantum jump every child must and usually does make. The freedom he had experienced living with his London cousins (they had been afraid he would seduce their daughter and gave him every reason to be out) exaggerated the slightest attempt at parental control until in the second week he felt stifled. The Grays were intelligent enough to realize this, and with a rush of resignation, let go.

For the first three weeks he never touched a racquet. Jag helped with the heavy farming and discovered poetry: Dickinson and Browning, strangely enough. He would readily admit that he didn't understand most of the phrases, but he enjoyed the feeling of it and the sound of the words—the real intention of poetry in the first place. Daily he disappeared, riding into the moors to read and think, never getting as deep as he might have, but adding a reinforcement to the necessary spiritual dimension of a personality. He didn't exactly find God or the secret of Truth, but he did stop jerking off.

Knowing when to take a break from pressure, and doing it, set a major and intelligent precedent for Jaguar. All circuit players and, by extension, anyone in any walk of life where artificial crises rise to staggering proportions must know when the moment has come to blow the fuse and go someplace, each to his own, the right environment that fits the need. Jaguar established in his mind that his home, the farm, was his unwinding station, and Miriam Gray was pleased, very pleased. He would always go there in the future.

When at last he was ready to pick up a racquet again, he started working hard, serving baskets of balls for hours, skipping rope and doing wind-sprints. With his father's help, he built a backboard on the back of the barn and tormented the cows' sleep with incessant pounding against it into the cool, farm-fragrant nights. When he practiced his service at the girl's school (he had to buy his own net), it was like summer reversed, that is, the patina was that of a summer resort deserted in autumn. The dormitory was empty. The girls in his repertory company had graduated; Sirena was married and living in Wales. The place was like a cemetery, and the sound of the balls had a brisk, lonely message. The echoes of the squeals and girlish laughter were still there, phantoms, more felt than heard.

By September, Jaguar was supercharged, horny, and ready for bear. He told his father of his decision not to return to school—to play tennis full time and get on with it.

". . . it seems to be what I do best. Guess I'll not be a history don and I can count to four figures, which is all the math I need. The English language is a mystery to me—foreign ones are gibberish. I reckon I'll go and do my thing, as they say up there, and make some money."

His father tapped his unlit pipe and slowly nodded. He liked the boy as well as loving him, flaws and all, and knew that the boy was right. Christ knows, Oxford was out of the question.

"Aye, lad, yes—I agree that you're making the right choice. I don't know where you came upon that tennis talent. It's the same type of blind instinct that designs the architecture of the honeycomb. Do it—and God bless. Do you need any money?"

"I do that, Dad."

"Right." Mr. Gray passed him an anticipated envelope of bills. "You can pay me back next year."

"I think I'll be able to by then. For interest, I'm going to throw in some big rollers, so you can push this house out behind the barn."

"How did you know?"

"Well, we seem to agree on most things. I've never understood quite what it's doing here—it looks weird."

"Don't say anything. Your mother loves it."

"No she doesn't, Dad. She told me last year it was a mistake, but you were used to it now."

"We'll see," his father said, deeply pleased.

When they all said their good-byes this time it wasn't a boy going away, but a young man; it wasn't quite as painful. Miriam could still see traces of her little boy in him, but only traces. He was a new, nice person who loved them and also happened to be their son. Could the time—could her life—have gone with such dizzying swiftness?

His return couldn't have been better timed as far as Ferdy was concerned. A place in the King's Cup squad had come open—Buster Mathews had bronchitis. It was a weak team at best. Jaguar demurred, saying that he wasn't match-hard yet, but Ferdy poo-pooed it. They had a week. Ferdy himself was leaving at Christmas; his contract was up and he was returning to Devon. Ferdy Archer wanted to go out a winner.

Jag didn't let him down; cleaned up his singles and was stellar in the doubles, and Britain pulled off a hat trick. Archer showed his gratitude by arranging for a chocolate company to sponsor Jag on the Caribbean circuit after the holidays. Jaguar promptly lost his sponsor by saying, on a television interview, "I don't eat sweets —the people I beat do. Sugar is poison." The company canceled three minutes later. Nevertheless, the LTA, a sporting goods firm, and, anonymously, Jag's dad came up with the money to sponsor him, and he counted the days, the moments until he could walk right into a giant travel brochure of paradise and sit down on the sand.

Bags and newly strung racquets were lined up in front of the scales and counter. Buster Mathews and Billy Sherman were in front of him, tickets in hand, checking and rechecking their new passports. Two hours later, they were casting winged shadows on the clouds over Shannon, homing like a vast missile to the lands of sand, sun, palms, and pussy.

The Bridgetown, Barbados, Air Terminal was aswirl with color and music. Veteran players were waving and smiling to one another as they hurried to taxis and waiting cars, while calypso smoothed the way into the warm, tropical evening. The young novice players, tyros on the tour, were looking at the famous players with something approaching awe tinged with nervousness, and at one another like strange dogs on unfamiliar territory.

Russell Kirkland, a once-great Jamaican player, was the promoter for this tour, and he left nothing to chance. He had made practice courts available, racquet stringers were on call, and the hospitality was warm and soothing.

Shep, fresh from a good NCAA showing in which he reached the semifinals, was met by a man with an identification badge stating simply *Tennis-Barney*. He ushered Shep to a waiting limousine, then returned to the terminal in search of his other charges. Within minutes, he returned with Jaguar Gray, Billy Sherman, and Buster Mathews, all looking bleary-eyed after an eleven-hour flight from England. They politely introduced themselves and lapsed into silence. Sherman lapsed into oblivion, with his head twisted at an awkward angle on the armrest. Barney disappeared once more.

As Shep silently appraised the English threesome, his confidence, a sometime thing, began to ebb. It was his first time out of the United States, and the unfamiliarity diluted his knowledge of his capabilities.

Jaguar was in shock, dazzled by the palm trees, the velvet, frangipani-filled night air, and the hubbub of the airport activity with all its strange flavors. He didn't intend to be rude. He was simply at that moment incapable of conversation—the introductions had been a painful effort in themselves. Mathews sat back with his eyes closed, humming along with calypso music.

Barney returned with young Al Wick and the group was complete. The car slipped into traffic, and Barney slipped into a monologue that he maintained without losing his breath for twenty minutes.

It was now dark. They were out of Bridgetown on their way to the Miramar Hotel, which was hosting them for the week (as long as they were still in the tournament). Wick, who had flown in from San Francisco and had been waiting for the English group's plane, was full of beer. He spoke only to make an occasional sarcastic aside, usually remarking on Barney's travelogue.

"I'm sure what you're telling us about is out there all right, but it's pitch dark, mate. What the fuck's the point?"

"Yes, mon, the point is that you can have the wonderment of discovering what I'm telling you about tomorrow in the light—instead of passing it in ignorance."

"Barney, your last name wouldn't be Gofer, would it?"

"No, my mon—a gopher is a small mammal—I'm a big bamboo." Barney laughed with a display of wonderful white teeth.

"Not that gopher. Gofer, as in, 'Barney, go fer this' and 'Barney, go fer that.' "

"You're beginning to vex me now. Let me get on with my explanation."

"Put a sock in it, Barney. You can tell us tomorrow when we aren't all tired. We'll appreciate it more," Shep mediated. Barney nodded and went quiet, and they all appreciated it more already.

With Barney's help, they were efficiently absorbed into the hotel and fell exhausted into their respective beds. Jaguar and Wick were thrown together in one room, the other two British boys shared one, and Shep was basking in the fact that Tom Gorman had agreed to share a room with him. Before the tour was over, Shep's starry-eyed impression of Gorman crystallized into the certainty that here was one of the nicest guys in tennis. They hit it off from the first moment and talked for an hour or so before dozing off.

Wick and Jaguar, demolished by travel and lack of sleep, had little to say, said it, and retired into themselves to listen to the tropical night sounds, the incessant roar of the ocean, and lonely, nocturnal birds. Jaguar was luxuriating in sense gratification and inhaled deeply of the wafting jasmine. When he was sure that Al Wick was asleep, he tiptoed over to the balcony to gaze out at the monochrome of the sporadically moonlit night. Wispy clouds were playing games with the moon, and the flashes of silver on the sea were of too short duration to fix a picture in the mind. Jaguar went back to bed and pulled the elements of the beauty he had seen into sleep with him.

The bright young faces at breakfast showed no signs of fatigue or jet lag. Some of the older players—Charlie Pasarell, Marty Reissen, Stolle, Panatta—still trying to get their eyes open over coffee or tea, remembered that it was only a short while ago that they could spring back like that, just a cough away in time. They were to have their revenge on the relative youth of their opponents later that day.

The beach was being pounded by waves and the feet of dozens of tennis players running. The locals gathered to watch the phenomenon of white maniacs running in the growing heat of the morning. The English group curtailed their laps. They really felt the heat, and although they were well-conditioned, were completely unprepared for the salt loss. The Australians were at home in it. Even Jesse Fraser, who had just flown in early that morning, was jogging on the hard sand.

Jaguar was doing reconnaissance on his laps and located and marked two groups, each with three lovely girls sitting on blankets. The hostesses, as they turned out to be, couldn't believe their luck when they beheld the sugarplum vision of so many handsome athletic bodies in motion in one place. When Jaguar, sweating and puffing, plopped down on the sand beside one of the blankets, asking all the dumb questions that young men ask to chat up girls, their layover had officially begun. After unloading their life credentials on the sand and speaking small nonsenses, it was arranged that Jaguar would leave complimentary tickets for them and celebrate his win or loss that night. His first match on the circuit. Then Jag stood up and, running flat out, assaulted a huge breaker—and lost. It curled him up on the beach after a few sandy somersaults. But the water was—warm! It actually was. He called Mathews and Billy Sherman to join him. They dove and larked about without actually swimming, which turns a tennis player's muscles into spaghetti. The girls joined them and they played all the sexual-overtoned water games until it was time for lunch.

The veteran players had, after their jog and plunge, showered and dressed for the morning's drill. They scattered by courtesy cars to the various courts made available to them. Jesse and Wick, borne along by the guidance of the older Australian players, John Alexander and Colin Dibley, left the beach to practice. Jaguar wrongly figured that, since he had a late-afternoon match, he could have a knock later to orient himself. The other two—Buster and Billy—followed his lead.

Perhaps it wouldn't have mattered much anyway, but Jaguar played at three o'clock and was annihilated by Gorman, winning only nine points in the first set and one game in the second. Without losing his charm, Gorman injected a seriousness into the game that, along with the acrobatic moves in change of direction that made him seem to be everywhere, completely undid Jag. As they shook hands, Jag said, managing a little smile to reflect Tom's Irish grin, "Thanks for the lesson." When he left the court, he found that his young colleagues had already departed for the hotel, befogged in their own dejection. "Could they have done any worse?" he wondered.

Shep, after *his* loss, didn't join the American contingent, but sat quietly in the players' enclosure, going over the match in his memory. He was still a very private person. An envelope with his hundred dollars was dropped on his lap by a tournament secretary. He stared at it for a long time, took a deep breath, and put it in his pocket, wondering how much the big boys were getting under the table.

Wick had been bulldozed by Ron Holmberg, whose elegance of stroke hypnotized him, lulled him into misreading the pace, and his points were few. He lost 2 and 2. He saw Jaguar and called to him. "How did you make out, pommie?"

"I earned my courtesy game. You?"

"Back to the drawing board. Are you heading on up to the hotel soon?"

"Whenever you're ready."

At that moment the stewardesses discovered them. "We've been looking for you all over the place," they chorused.

"I was in the bar consoling myself," Wick answered.

"Girls, this is Al Wick," Jaguar said. They all shook hands and gave their names, which faded before reaching Wick's ears, it seemed to him.

"We thought you both played terrific," one of the stewardesses said.

Wick and Jag shuffled uneasily in embarrassment. "Sure," Wick said, "I can't wait to go public so I can buy shares in myself. Let's face it, I took a hiding out there."

The girls were still enthusiastic over their straw heroes. "But you hit some good ones," another mollified.

"Great! Like pissing in a cyclone." Wick laughed. "That fucking Jesse won his match—oh, sorry, girls." They didn't quaver. It was a good test.

When they had found their driver and climbed into the car, Shep ran up to stop them. "Room for one more?"

They looked around reluctantly. "I suppose we could squeeze you in if the girls don't mind sitting on our laps," Jaguar decided, brightening to the idea. Wick knew that the novelty of having a hundred-and-ten-pound girl on his lap would wear off after the first ten minutes, leaving him semiparalyzed.

"Sorry about this," Shep said. "I can't find the guy who's supposed to drive me and I want to hit some balls. Are any of you interested?" They agreed to play threes on the hotel courts if they weren't booked.

"Did anyone our age get through?" Shep wanted to know.

"Only Jesse," said Wick. "Lucky bastard played a local and had a nice win. But tomorrow he plays Laver. That should be pretty funny. What happens if we lose our doubles tomorrow? Do we have to leave the hotel?"

"Al, it's that or pay fifty dollars a day, which is half a week's pay if you had the same thing in your paycheck." Shep smiled at the hypocrisy of it all. Amateurs indeed.

"It's fair enough, I reckon. Gives extra incentive to win the doubles."

Shep and Wick were talking underneath the girls' conversation, which was on another plane.

Jaguar's conversation had reached the whispering stage in the front seat. His lady jockey had a hold on his tool, and the driver was trying to peek and watch the road at the same time. There were several near misses, filling the car with shrieks of mirth and roller-coaster laughter. At one point, Wick asked to take the wheel, but the driver wasn't giving up his seat for anything. The hostess in front was experienced enough to know what she had in her hand and wasn't about to let go of it.

They left the girls with a plan to meet for dinner (dutch) and went up to the courts to mend their wounds. There was a waiting line: Billy Sherman and Buster Mathews had one; John Newcombe and Tony Roche, no less, were whacking scorchers on the other. A few players waited for one, and a group of tourists were watching. Most of them were delighted to be able to see the "boys" doing it right at close hand, but there were exceptions, a few surly ones. One American kept muttering that this was his first holiday in three years and, paying a king's ransom, he still couldn't get on a court.

Shep overheard him and invited him with his wife to share the next free court. They accepted it sourly, with a great display of loosening-up activity. They were playing with nobodies, young kids, that day, but to hear them tell it a few year's later—oh, yes, they knew potential genius when they saw it.

After a swim, hot tub, shower, and dinner, the group met in the suite the sky-ladies were sharing, for a party, the results of which were a foregone conclusion.

It didn't start out as an orgy. Wick arrived with a three-piece steel band he had found along the road. Shep came in with four American secretaries he had just met in the elevator. They were cautious but game and, after a few minutes of clinging together for mutual security, they were infiltrated by Shep, Wick, and a young, aristrocratic Frenchman, Jean-Claude Dumée. Although effete looking, he was exceedingly attractive to the girls, and his *fin bec* wasn't too *fin* for a bit of hanky-panky.

Rum, gin, and beer appeared magically with the compliments of the house, and a full-scale brawl was born. Players and hotel guests drifted in and out through the evening, the husbands using their wives as entree, a joining fee. The dancing wasn't authentic merengué, but they enjoyed it at least as much as if it were. Even Jesse dropped by for a half hour and drank a beer. The first mouthful gagged him. He kept remembering Laver, probably asleep somewhere already. Jesse slipped away and went to bed. He might as well have stayed at the party for all the sleep he found. At about one in the morning, the musicians, after a note from the management, took their quiet leave.

Jaguar sequestered himself and his young lovely in one of the three bedrooms. After a decent interval, Billy and young Mathews did the same, whipping off with their dinner companions.

Two of the secretaries were down to bra and panties, improvising to Jean-Claude's French interpretation of a Leadbelly number. He took a nap in the middle of a bar, and they continued without him, removing the little wisps of clothing left and tiptoeing like wood nymphs around the room, teetering here and there, all the while giggling and balancing great glasses of molasses-tasting loud-mouth juice.

Shep brought one of the girls back from his room and, feeling delightedly decadent, curled up on the sofa with a second wave of plenty, a naked lady who alternately sang to him and kissed him. Jaguar appeared from the bedroom wrapped in a huge towel and ordered whipped cream from room service. When he had hung up the phone, he turned to Shep with a shrug. "She needs it for something."

Several of the veterans made a flying raid and spirited off three of the American girls, including a nude and two still-lifes.

Jaguar came out again a bit later for the rum bottle and stage-whispered, "She did a number on me with her mouth full of hot water . . . and you wouldn't believe what we did next." Gone again.

The deep, dark tan of Shep's dancing partner emphasized the

small white band that girded her, lending to her the illusion of seeming to be dressed, garnished with a dark cleft behind and an ornate, glistening, golden bush. She said softly in Shep's ear, "Do you want me to show you what they're doing next? Come on, Sheppy Wheppy. Come on and I'll show you." They ducked out of the party, which now consisted only of Wick and a young newlywed couple. Wick was telling jokes again. "Little boy asks his father, 'What's a lesbian, Daddy?' His dad answers, 'Ask your mother, he'll tell you.' "

Shep laughed as they dogtrotted down the hall, partly at Wick's joke and partly at the spectacle he and his lady friend made together; he in blue blazer and red-and-white candy-striped undershorts; she, stark naked. They passed an elderly couple in the hall. The old man acquired a new jab in the ribs from his wife's elbow when he tried to look around. And the ocean kept pounding the shore as if nothing were happening.

Morning was its own retribution. People emerged from their rooms blinking like mice. The hotel was a vast, throbbing hangover. Shep, who had stuck to beer, was in reasonable shape after he threw up. But those who had soaked up the rum participated in a painful awakening in the full. Jaguar bounced off two walls, spun around, and fell back on the bed. His head was splitting with pulsing agony. The pile of sheets next to him stirred. Slowly a hand reached out and stroked his dormant member. He pushed it away. Hair of the dog, maybe, but the necessary motions for hair of the girl were far too nauseating to even consider.

The gin twins, Mathews and Sherman, crept out of their respective love nests, grimaced at each other as they met, and went down the hall to their room for rest and recuperation. For many years afterward they would remember the piercing icepick-in-the-ear sound of a yellow bird somewhere chirping viciously away.

Jean-Claude still looked *raffiné* and intact, except for undeniable evidence to the contrary. He was lying in the shower fully clothed and it was running. He was wondering how the top players could dissipate like that every night and still function. The answer, naturally, was that they didn't.

On court, Jesse kept solacing himself with what an honor it was to play the great "Rocket," as he took the worst beating of his young career. His best shots came back with dividends. Of course, he *was* immensely intimidated by the fact that he was playing Laver, but, regardless, Rodney was bringing out Jess's best tennis and still doing a housewrecking job on him. He lost 6-1, 6-1, and for years wondered how he had managed to win the two games.

By a coincidence of the draw, Al Wick and Jesse were playing Jag and Billy Sherman that same afternoon. Since his match had taken only forty minutes, Jesse was still fresh . . . but Wick kept belching raw rum and saying disparaging things about himself out loud. Four of them wanted to win, but three were as glad just to get it over with. Wick said, when it was finished and he and Jesse, no thanks to him, had won: "My mother used to tell me that masturbation was the murder of a perfectly good emotion. I feel as though my whole body's been masturbated. I'm going back and lie on the beach until I get well."

He wasn't the only one with that solution. The beach in front of the Miramar was sprinkled with inert bodies among the coconut trees. A rogue breeze would occasionally flutter the hair on some of them, but otherwise there was almost no movement.

By dinnertime they had, for the most part, recovered enough to privately swear off parties and make their reservations for Kingston, Jamaica, on the next leg of the tour. It was an empty feeling, being around a tournament when one wasn't participating in it any longer. Most of them were going to leave, but Jesse and Wick were still in the doubles: they were to play Ralston and Ashe the next day. "Barring a miracle, we'll see you tomorrow night or early the next day," Jesse said. They were going to give it everything they had, but one must be realistic.

"I'm not letting you out of my sight, Wicko," Jesse warned. "Starting right now. Two beers and off to bed. I've been moved over to your room, since Jaguar is leaving tonight. At ten, I'm turning out the light, so don't play your flaming transistor radio." He turned to the others. "He's like a kid with a new Christmas toy."

"Jesse doesn't like classical music." Wick defended himself. "If it was country and western, he'd be as happy as a pig in shit." The Australian Odd Couple.

During the ensuing lull, a night bird broke the silence. They looked up from their food. It sounded like a flute, rich and resonant, with glittering runs. It gave them all a momentary lift of spirits.

Jaguar usually had a mean appetite, but tonight the beef stared up at him in clotted disdain. "I have to finish packing and catch a plane. Come on, Billy, you're in charge of getting Buster up and ready in fifteen minutes. I'm not waiting for you if you aren't. I've turned into a flaming nanny, that's what." He waved to the group. "See you tomorrow, or whenever."

11

Kingston was a sprawling city at the foot of the Blue Mountains. The driver, with great pride, told them that Kingston's harbor was natural and twenty square miles in area. They drove around part of it coming in from the airport. The blend of shuttered old buildings and skyscrapers was in sharp contrast to Bridgetown's architecture. The narrow downtown streets broadened into avenues as they hit the suburbs. The driver went on about Port Royal and the pirates, the revolutions, the national heroes, Bogle and Garvey, about Henry Morgan, who was Captain Blood in Errol Flynn's great film, and on and on and on. The driver was named Louie, but to the British group he was Barney. After a while, he just shrugged and *was* Barney. What the hell.

Billy, sitting up front, was turned around in the seat speaking to the others. "Hey, Jag, do you think they'll give us the golden handshake tonight? I'm a bit short of the old, ready whip-out."

Jaguar was watching the city disappearing behind them as they turned up a road through a forest of sugar cane. "Now how the hell would I know?"

Billy turned around and was silent. Buster Mathews hadn't said anything for hours but sat brooding.

"Buster here hasn't said a dicky bird all night. What's the matter?" Jaguar asked Sherman.

"That airplane person last night . . . fell asleep before he could put it to her. He didn't know what to do, so he went to sleep too. Now it's annoying him that he missed the chance."

"Cheer up, Buster. There are more in the sky where she came from," Jaguar said, with just a hint of mocking in his voice.

Buster finally broke his vow of silence. "All right for you to say, Jag. You're getting enough crumpet to wallop a dog out of a sausage fact'ry."

"Cheeky monkey," Jag said, imitating a British comedian.

"Why didn't you just give her a poke anyway, Buster?" Sherman asked.

"Yeah . . . I should have. It just didn't seem proper at the time, somehow."

They pulled up in front of an old colonial-looking building on what seemed to be a private estate. It was slightly seamy, but not quite run-down. A little paint and fixing and it would be splendid —a little less of each and it would be ramshackle. It was a former manor of the cane plantation, and was now a sometime hotel and residency.

Madeline, a lovely, thirty-year-old café-au-lait in a twenty-year-old body, greeted them. She was a cousin of Russell Kirkland, the promoter of the Caribbean tour, and was acting as hostess for the eight people who would be staying there for two days until hospitality could be arranged for them closer to the town and the tournament. She took one full blast from Jaguar's vitreous, pale blue eyes and made an instant decision. She chose well. Jaguar's experience was gathering direction like a downhill racer who learns the little moves to cope with speed. He was a quick study.

By now it was late and, after sandwiches of cold cuts and iced tea, they retired. Madeline gave Jag's hand that extra squeeze in saying good night; he nodded. Jag was a trouper now. He knew the scenarios and could arrive almost anywhere saying, "Where's the stage and what's the play?"

He had almost dozed off when the light rap at the door made an educated guess a fact. He let her in with stealth—simply out of habit—not knowing if it were actually required or not. She ran her long, slender fingers down his cheeks and looked hotly at him with her eyes—brown, bottomless, liquid, all-knowing eyes. They embraced in a full-tongued kiss. "I bet she could suck the chrome off a trailer hitch," Jaguar thought.

Madeline disengaged herself, lit a kerosene lamp, and turned off the lights. There was a mosquito net, but the ancient overhead fan was enough to keep the bugs off so they didn't pull it down. During the foreplay, Jag's southbound lips and tongue hesitated at her thick, perfumed bush. Then he changed his mind and decided that he was too hot to wait, too anxious to discover if he was right about her virtuosity with those full lips. He wanted her to go first, but she liked his initial idea better and resisted the slight pressure he put on her head, saying with phoney naïveté, "But I never did that—I don't know how to do that."

"What's to know? Don't bite and don't spill a drop." His presentation was too impatient, almost belligerent, with a selfish ring to it. Madeline pulled away from him angrily, grabbed her clothes, and disappeared through a connecting door into the next room. Hers, Jag reckoned.

Jaguar clucked his tongue a few times, gave a win-some/lose-some shrug, and blew out the lamp. The day had been rich enough in sense experience as it was. He turned off the noisy old fan and dropped the mosquito net. The airless pale of night enveloped him with a sonic texture that perceptibly quivered. The tropics massaged him to sleep. His last thought was of England in January. Even half-asleep he shuddered.

Shep was mystified as they rolled up to the rococo house. In the bright Jamaican morning light, the house reflected not so much a former time of splendor and opulence as a shaky anachronism. Unlike the ersatz Barney who had mesmerized the English boys with his spiel, Shep's driver said nothing. It seemed all the world like a

Thunderball plot by S.M.E.R.S.H., and Francis Shepard jokingly wondered aloud why he was being kidnapped.

The car rolled up to the vast verandah, and Shep climbed out in sections as if his limbs were articulated. Madeline received him with a false gushiness which was, to the practiced eye, the behavior of a woman in a very bad humor. When he saw Jaguar around the corner rocking feverishly in an old wicker chair, he knew he'd guessed the source of her ill humor. "Hey, tiger. Take a midnight ride to change your luck?"

"Not I, mate . . . never left my bed." By the nature of the answer, Shep was satisfied that his surmise had been correct.

Billy appeared on the porch and, after a perfunctory wave to Shep, asked, "What's the drill here? Do we get to practice or what?"

The question was coincidentally answered by Madeline, who had materialized on the porch as a splash, a blur of pink and creamy chocolate. She ignored Jaguar and spoke directly to Shep. "We've resurrected the court for you. I think it's all right. Now, the car won't be back for about two hours with the others, so if you want the court to yourselves . . . or better yet, if you want to swim, Harry will show you where the river basin is behind the mango grove. We also have three horses for you, if anyone cares to ride. Lunch baskets are in the kitchen made up. I have to see about tonight's dinner, but Harry will be able to help you . . . to answer your questions. Won't you Harry?"

Shep had been to a lesbian club in Bridgetown a couple of nights before called Dirty Harry's Bar, and Madeline's major-domo was a dead ringer for him. Jaguar convinced Billy to go riding with him. They would swim on the way back. Shep didn't want any new aches and pains, so he demurred. Buster, whose 6'4", 155 pounds (the "Jolly Green Stringbean") made him look enough like Ichabod Crane without getting on a horse, hated horses. He was convinced that God hadn't created man to wrap his legs around a giant beast. Horses didn't like Buster either.

The boys grabbed their baskets and headed outdoors. Shep decided, on second thought, that a gentle ride as far as the river wouldn't put his pelvic muscles too much out of whack. He mounted a small bay. Buster walked alongside and, often, ahead of the three horses. The trail led through a small mango grove with gorgeous doctor birds warbling away, then into a lush, tropical tract with thousands of greens, festooning bougainvillea, and wild orchids, the dappled sunlight filtering through the high tangle. To Jaguar and the other British boys, it was a fairyland; to Shep, it was as wonderful as he had imagined it, which was exceedingly high praise.

They reached a crystal-clear stream and followed it for fifty yards until it became a small but heavy cascade falling into a lime-blue basin. Jaguar crossed the stream above the falls, dismounted, and took a picture of the scene.

"Come on, Billy!" he called.

Billy's legs were already straining around the big mare he was riding. "How about a swim first?" he called from the opposite bank.

"Let's take one when we get back. If we go now, we can be back before it's too hot." Jaguar walked off into the dark green foliage. Sherman urged his horse into the stream resignedly, squirming in the saddle, and followed Jaguar into the forest.

Shep and Buster stayed behind at the falls. This was Shep's only other tournament on the Caribbean circuit. He had to be back at school the middle of the following week. Shep knew that he would probably make school on time. If a miracle happened and he did get to the third round of singles or doubles, it would be well worth being a little late. His grades were good and his tennis position was solid. So he gave himself up to the beauty of the moment, almost approaching Jaguar's constant sense-state of spongelike absorption. Shep and Buster stepped through the falls into a cave beyond, and sat there viewing the world darkly through a silver screen. Buster was a shy, taciturn boy and they were engulfed in their own thoughts. Finally, Buster said, "Sort of like being in a watery cage in here."

"We all live in some type of cage," Shep answered dreamily. "Some people's cages are bigger than others, that's all."

Buster thought about that for a while. Then he stripped and, with a gangling leap, dove through his cage into the pool. Shep followed him. The water was chilly but glorious. After a brisk swim, they sat on a big rock and ate their lunches. When their saturation point of beauty had been reached, they returned to the house to practice.

The court wast in first-rate condition: resurfaced, rolled, and frequently watered. Every fifteen minutes, young Jamaicans swept the court, wet it down, and cleaned the lines while Shep and Buster drank lemonade and swallowed salt pills. Shep was used to the heat rising hungrily off the concrete in California, but Buster still couldn't hack it.

About the time they were packing it in, Jag and Sherman reined in their horses after a brief canter on the home stretch. Billy crawled off his mount with painful, precise, angry movements. "Fuck you, Jag, for a beautiful morning. I'll never walk again." He was in agony and flopped down on the grass beside the court.

"Go take a very hot tub right away, and then come out and hit," Jaguar told him. "If you let the muscles completely stiffen up, well, you've had it, mate, haven't you?" As usual, Billy did as he was told.

Later, on the verandah, which had become headquarters, the four of them sat under giant fans with their tall, frosted drinks. Shep was still something of a stranger to them, with his American ways, but a shared orgy and being thrown together in an exotic, neutral surrounding had broken most of the ice. They chatted lightly and aimlessly. Most of the conversation centered around famous players. Shep was going on about Ion Tiriac, the Romanian stalwart whose exploits and histrionics had made him a walking legend: his fierce clowning, the feat of eating martini glasses, stems and all, made him seem a sophisticated barbarian. "He was so mean he could peel a potato at twenty yards by hating it," Shep said with a laugh.

The restful, indolent mood of semi-reverie was broken when a car pulled up with four passengers. Wick got out and held the door for Frances Ferris, a player in her twilight years of top tennis. That was a plus, having a second girl there. Jesse climbed out, balancing his racquets.

"How'd ya do with the Rocket?" Billy called from the porch.

In answer, Jesse rolled his eyes and threw up his hands in mock horror.

When Randy Mariano emerged there was an audible "Oh, shit!" from both Shep and Jaguar. Shep turned his head to stare quizzically at Jag, then recalled that Jaguar had beaten Mariano in the Junior Wimbledon that Shep should have played two years before. "That bloke's a right bastard," Jaguar whispered.

"I know, I know," Shep answered.

Jesse and Wick had their Australian Davis Cup blazers on over green Lacoste shirts, and they were sweating in visible rivulets. They had anticipated a short ride from the airport and had wanted to make a good impression.

"You're both looking sartorially splendid on this fine, hot day," Shep said with his painful version of an English accent.

"You've been reading again," Wick shot back with some agitation. His armpits were dark green with perspiration. The staff rushed down for the bags.

Mariano stretched and looked around. "Hey, man . . . what's happening here? Are you doing a remake of *The Fall of the House of Usher?* That house must be a thousand years old."

"Welcome to our humble home," said Billy, with a short bow.

"Yeah," Wick observed, "don't knock it. It might fall down."

"It's only paradise, you fucking cynics. Don't spoil it for the rest of us." Everyone turned to stare at Buster, who had just stepped out of character with that statement and just as quickly hopped back in—to silence.

"Gentlemen, really!" Frances feigned shock in a phoney Southern accent. "What language! What manners!" She took Shep's arm on the stairs and sashayed like a great lady up and across the verandah. "Thank you," she said with a slight curtsy. "I shall have a mint julep and relax while you gentlemen come to your senses."

"What a dump! Why are we way the hell out here? Don't the darkies want us in their homes?" Mariano asked. His lip was slightly distorted with sullenness.

"I'm a darkie, and for the moment, this *is* my home," Madeline said icily. Randy's condescending toss of his head did nothing to mollify the moment.

"Shut your mouth, you clot," Wick said aside to him, "or I'll kick your ass up between your shoulder blades."

"Jesus Christ!" Shep whispered to no one in particular. "He's only been here one minute and it's already paradise lost." Madeline stayed to supervise the handling of the luggage. Shep led the way inside with Frances on his arm.

The rest of the afternoon was spent practicing. They broke up into two groups of four, alternating hours. The last two hours of light, Jesse and Wick played a practice set, then Jaguar and Shep practiced lobbing and smashing. Billy Sherman was crippled; Buster, exhausted; Frances was swimming; and Mariano, uninvited.

Dinner was a candlelight masterpiece of Jamaican cooking. Madeline sat at the head of the table of glittering silver and multihued floral centerpieces, telling amusing and spellbinding tales about Jamaican folklore, especially about the Maroons who lived in the mountains. "They were wild, exciting people, all right. Oh, yes, their name. The word was derived, first from the Spanish 'Cimaron,' meaning escapee. Then it was corrupted to the French 'Marron,' meaning the same thing, and finally, Maroon, the slaves who had escaped into the hills and mounted a revolt."

Jaguar was conciliatory in his manner toward her and she gave him tacit reason to believe that the night before had never happened.

Mariano heeded Wick's warning and was almost mute the whole meal. He wasn't afraid of Wick. Randy was a good street fighter. He had to be. But he also knew that he couldn't bluff Wick and that he'd probably have to kill him to win. These assholes weren't worth the grief anyway, in his consideration. He opted for silence. So the dinner was delightful in spite of him.

When Madeline knocked that night, Jaguar was no longer cavalier; he had grown up a little. It was well worth the change in attitude. In the other wing, Shep and Frances glided into sleep together after some friendly, non-acrobatic sex.

The first morning light found them hitting balls at a hot pace in the lush cool of the green-enshrouded court. They changed off all around so that everyone had an opportunity—two balls going continuously. After a short, brisk, bracing swim, good-byes and well-wishing were exchanged, and they joined the luggage in two waiting cars. The group was to be parceled out individually among middle- and upper-class families in Kingston. It had been fun being together in the little shambles of paradise. Mariano, on Madeline's instructions, had gone off into the hills to find a cockfight and was gone the better part of the morning. He didn't find the fight and missed the ride in, having to wait for the provision truck which made a run later in the day. Randy was very put out. None of the other staff had ever heard of a cockfight in the morning.

It shouldn't have been such a surprise to Shep that middle-class black families lived normal, loving lives and weren't Amos and Andys or Catfish Row showboaters. Charlie Lumsdale was an orthodontist with a flourishing, if exhausting, practice who played tennis, was a fanatic about cricket, and had little time for either. He had canceled one appointment in order to meet Shep and make him feel at home, but had to hurry back for the next one. Shep's impression of him was that of a gentle, handsome, early-middle-aged man with an omnipresent sense of humor and a passion to provide for his wife and three children. Mrs. Lumsdale worked as a researcher in a British Assurance branch, and the children, a boy and two girls, were all of school age. The twelve-year-old boy had good manners and was mad for tennis. He was content to sit with Shep's racquets on his lap and dream. The girls were squeaky shy at first, then all love and attention. Nothing was forced. Everything—manners, attitudes, and friendliness—flowed easily, and Shep felt, very soon, a part of their lives. Naturally, he didn't see them much in the daytime—an hour at breakfast, a fast set with Charlie one afternoon, and the children's hour in the early evenings, but Shep had entirely forgotten about color. The one moment when he was aware of it and surprised to discover once again that they were black, he hoped that they too had forgotten he was white. In a few short days he came to feel very close to them and to care about the children's school work and dreams of the future. He gave the boy, Sydney, a new racquet and Sydney embarrassed himself by choking up with tears.

Charlie never told jokes. He just said very funny things. His accent wasn't a pronounced Jamaican, but he could shift into it when it served his purpose, and when he did he cracked Shep up.

Sydney became Shep's official ball boy for practice sessions and it was arranged for him to be enrolled in the next ball boy classes at the tennis club. Shep hoped that they liked him as much as he them; he could only speculate on that, but he consciously kept from being corny and overly demonstrative in order to avoid giving a condescending impression—the big, kindly white man and the cute pickaninnies rubbish. He knew the name of that tune is bullshit. He shuddered to think who had inherited Mariano. Randy wasn't a racist—he just didn't dislike anyone more than anybody else.

On Monday, Shep gave them all something to be proud of by drubbing another young man, a player from Montego Bay. After the match, the young loser gave him such a glowing description of

Mo' Bay, as he called it, that Shep half decided to make a stop there on the way home.

"I want to go too," Sydney said, with a desperate quality in his voice. "I don't have any school the last days of the week. I could visit Perry."

"Sydney, my little mon, this handsome young fellow doesn't want kids traipsing along in his life and being sticky around him."

"I'm not looking for action, Charlie," said Shep. "If I do go, I wouldn't mind at all if Sydney came along." Given Shep's nihilistic frame of mind in those days, living with the Lumsdales was a rich experience, an enlightening boost that strengthened and broadened his character.

Jaguar came off court, having just won his first match on the circuit, but when he had Shep alone for a minute, his conversation wasn't about tennis but about the unique experience he was undergoing with the black people who were putting him up. "They're bloody marvelous. Just like white people only they have more fun."

Shep nodded, understanding completely, and more than just a little amused at Jaguar's brave new vision. Jaguar went on: "While I was checking on my doubles in the referee's office, Mariano was in there. He's a bit outrageous, he is. Just won his match at love, love, and scratched. He said he didn't like his accommodations. He stressed that *'accommodations,'* and if they wouldn't put him up at a hotel . . . well . . . he couldn't afford it. Took his first-round money and scarpered. He's probably on his way to the airport by now."

"I'd like to know the full story. I sure hope he didn't hurt those folks' feelings . . . but I guess the odds are that he did."

"He's a real shit!"

"Number one . . . a cruiser-class shit. He dropped out of school to be full time at it."

Jesse and Wick were talking by the door. When they spotted Jaguar, Wick called, "Hey, Jag, how're you doing?"

"Not enough, old boy," Jag answered in a simulated West End accent.

"Well, for Christ's sake, don't slow down. Who else will help take up the slack. Someone has to do our share."

"How about this dirty sheep dog here?"

"Nah . . . Americans are too fussy to fool with our stuff."

"No I'm not. I never screwed a girl I didn't like."

"Catchy. How are you doing otherwise, Jag?"

"We've both managed to scrape home, thank you," Jaguar said with undisguised happiness.

"Right . . . well done. Who's next?"

"Jaguar's going up against Fred Stolle, and I have the honor of having my all-American ass waxed by Laver," Shep said as he cleared away room on the table for the Lumsdales to put their lunch trays down.

"Beautiful, beautiful, Shep, love . . . give him hell. Play his backhand . . . it's his weakness." They all laughed and went into their separate worlds of the moment.

The second rounds all went as expected: no surprises, no shocking upsets. Shep no longer had to worry about getting back to school on time. Laver made it possible for him to have the rest of the tournament off. If not for the few kind words the "Rocket" said to him about his potential, Shep would have been a thoroughly demoralized boy. Laver took his game apart brutally. With apologies to Sydney, Shep flew straight back to California.

With their inevitable losses in the next rounds, the young group, with sincere thanks to their hosts, preceded the rest of the show on to the next island, Puerto Rico. They were meshing into a life-form which was to be theirs for the next few years. Hotels or new families, early losses, short purses, identical-looking planes and airports, mimeograph romances, repetitive conversations and on and on. . . . At first, it was refreshing, but it soon became a movie seen ten times, the senses bludgeoned by the duplications and numbed by a lack of real differences. They became insulated, traveling through the various cities and nations wrapped in cellophane; like an English Channel swimmer . . . oiled from head to foot . . swimming thirty miles and never getting wet.

12

Russell Kirkland's associate had a Puerto Rican "Barney" to meet them and take them to a new, artfully conceived hotel on the ocean just outside the lovely but incongruous city of San Juan with its jarring concrete presence soaking up sunshine and re-radiating the heat between the buildings. Later, when Jaguar saw Miami for the first time, he was struck by the resemblance. The San Juan leg of the tour followed the pattern: plane—check into hotel—reorient—attempt to familiarize with the slightly different surface.

Jean-Claude Dumée, Jaguar's new partner in crime, had arrived before them and had a small group of honeys primed and prepped for the arriving surge of the "boys." Except for a few matinees, it was agreed not to have any parties until they were out of the tournament entirely. Jesse was off women for the "duration," after Wick had dragged him to a whorehouse in Kingston called the

Oriental Laundry. Jesse remembered his cajoling words: "Come on, let's go down and reorient some Chinese pussy." For Jesse's puritan ethic, it had been a nightmare. Wick seemed to enjoy it all right and was very entertaining in spite of his disappointment in finding that there weren't any Chinese girls. The squalid aspects of the place were enough to put Jesse on the straight and narrow. He made many long-standing resolutions.

Only two of the girls in the ensemble of lovelies that Jean-Claude had mustered in San Juan were pros who had infiltrated the amateurs: secretaries, stewardesses, divorcées, and locals made up the "instant groupie" contingent. One of the hookers was Chinese, which delighted Wick. Alas, she was smitten by Jaguar. Wick said, "I should have guessed it. This morning a mosquito landed on my arm—took a closer look and flew away. That's real rejection. Like the man says, I don't get no respect. What was his line again? Oh . . . right. His wife says to him, 'Go on and take the garbage out.' He says, 'I did already.' And she says 'Well, go keep an eye on it.' Never mind, Jag. The best of British luck to you."

"Go ahead and take her, mate, but she's a very proper Chinese girl. Doesn't smoke or talk dirty. She speak nice . . . she call my John Thomas 'Charlie.' She say, 'Charlie want a kiss?' I answer, 'I'll get him on the hot line, but he's never turned one down yet.' "

"Fuck you, Jago . . . I hope sincerely that she gives you a package of something."

"Don't even mention that. It's not a funny subject." Jaguar had gone quite serious.

"Keep it in mind, Jago. See yah."

The young rabbits all lost on time and according to script, had their non-party, which, outside of a lot of meaningless screwing, was a bomb, and prepared to leave for Haiti on the last leg of the Carib tour. Jaguar was on his way around the royal-blue pool toward the office to check out when his eye spotted something and he went stone-struck-riveted to the spot. A beautiful young woman was standing inside the partially open sliding door of her room, naked Her hand held her bushy triangle and she had one hell of a dreamy expression on her face. Jaguar took a step toward the doorway. She didn't move. His heart struck up the old familiar beat and his throat constricted and his mouth went dry. He stepped slightly clumsily into the room. From the bathroom, he heard the shower running and a man singing *"La donna è mobile"* off key. Jaguar reached out to her and she walked into his outstretched arms. He kissed her hotly and ran his hands up and down her back, then gripped the cleft of her full bottom. As he reached down to hook a finger in the wet interstice between her legs, he glanced warily at the bathroom door. His heart was thumping. The running water and singing continued. Expertly undoing and dropping his trousers with one hand, Jag bent slightly at the knees and inserted himself. It was uncomfortable and hard on the thighs. Chain-stepping with his legs locked in his pants, he moved her backward onto the bed. The water stopped but Jaguar didn't. The singing stopped moments later. Jaguar was close. She was breathing like a bellows and murmuring the murmurs of the lost in passion. The bowl flushed just as Jaguar came. Without ceremony, he rolled off her onto the floor and under the bed in two deft moves—just as the door opened. From under the bed he saw a hairy, naked, heavily built man appear, drying his hair with a towel. The man seemed startled at what he saw on the bed.

"Have you been playing with yourself without me again?" he boomed with a thick New York accent. No answer. "Or worse!" He wrapped himself in the towel as he ran to the open door to look outside. Under the bed with Jaguar were two different socks, an airline-ticket folder, and a foul-smelling, full ashtray with a snapped popper capsule on top.

"What's been going on here?" At the sound of his voice, the woman came around suddenly and sprinted for the toilet, locking herself in. Jaguar watched her feet and ankles disappear. Then the hairy face mooned upside down at him.

"What are you doing there?" it growled.

Jaguar couldn't resist the stock answer. "Everybody has to be someplace," he said, with a philosophical smile.

The bare foot missed his face by inches and cracked a shin on the wooden baseboard of the bed. A shout of pain and garble of curses—then the man began to jump up and down on the bed. The sides caved in and the springs made little cookie patterns on Jaguar's face. Timing one of the jumps, Jag rolled out from underneath, tipped the man over backward with the mattress, and hobbled lock-step for the door and freedom. He wasn't followed. From the distance which he put between them once he hauled his trousers up, and hauled ass, he could hear shouting and a door being pounded.

On his way to the office, walking jauntily, he passed several of the hotel staff running toward the fracas. They all glanced strangely at his forehead, which was embossed with a big, red circle. His very own scarlet letter. Later that night, his story of being attacked by a mad midget with a toilet plunger didn't wash. Nor did the truth when he finally told it.

13

The Haitian "Barney Gofers" took some of the group to the Oloff-son, an architectural monstrosity which had been, successively, the president's mansion (late 1800s), a U.S. Marines hospital (1915), a hotel started by a Norwegian ship captain, then a French hotel, and now an American hotel. Each of the owners had added a bizarre touch to the incredible, rambling, gingerbread house. It did, as advertised, look like a Mississippi riverboat marooned on a Caribbean Island. To Jaguar and Jesse, it was, with its zigzagging stairways, filigrees, and towers, very reminiscent of the place in Jamaica, and they felt immediately at home.

Al Wick and the remainder of the English boys were billeted in the Royal Haitian Club Hotel, a new, unpretentious luxury hotel with tennis courts and good food. The casino of Port-au-Prince was the parent organization. It was owned by Mike Mclaney, who injected his charm and energies to make it more relaxed and friendly than the other stiff, impersonal gambling spas of the world. Most of the "high rollers" were set up in the hotel gratis. The junket covered it, so they weren't your normal guests.

Mariano was already there, having scratched early in Jamaica to get at the blackjack tables. He was sunning by the blue, breeze-rippled pool when the others checked in. They pointedly snubbed him except for Buster, who ingenuously nodded. Buster had other things on his mind. He had diarrhea, homesickness, and badly shaken confidence—he was taking tremendous shellackings at the hands of the pros. While the others checked in, Buster made a reservation back to England and a large family of disappointed Mathewses. He clearly wasn't ready for the big time.

Not all of the long knives were going to play the Haiti leg. Many of them were playing an ancillary, special series of matches in Vera Cruz, Mexico. When Wick found that out, he was very glad indeed. "That ought to soften them up. Couple of meals over there and they'll open up like a flower. From what I hear, the Mex-two-step really takes it out of you."

Refreshing showers and iced drinks cooled them down inside and out. Mclaney had been a class player in his day and knew the needs and temperament of the boys. While he made arrangements for them, they ordered creole omelettes for lunch and fell into speculations about the other poolside denizens of the shallow. Several elderly couples, going on their penultimate dogs, were trying to inject some warmth into their marrows. The remnants of a cruise-ship landing party were evaporating like tiny puddles on a sunny day—their weird attempt at holiday and shipboard attire made them appear like "Let's Make a Deal" contestants or losers in a Hadassah costume party. The last of them, a group of about twenty, were sucked through the lobby door by a vacuum in the bus.

The gamblers looked out of place in the sunshine. The winners were lunching or boozing under cheerful green and white umbrellas; the losers were lying immobile on deck chairs, with white plastic eye-guards on lotioned faces, or propped up like mutants, their one-way mirror glasses turning their eyes into private little rooms.

Why Mariano wanted to become a part of that insane, dyspeptic world baffled the rest of the boys, but not any more than grown men running around in shorts in the hot sun after a ball for peanuts baffled the gamblers.

The tournament, unfortunately for the organizers, fell in the middle of Mardi Gras, and the fever was building up already. It would cut into opening day and, of course, Tuesday, but those days were not especially big gate in any event, so it was a terrific plus for the "boys." The American Club at Petionville was to be the arena for the tourney. (Haitians hadn't been allowed to play there in previous years, but that policy had changed.)

The courts were in wonderful condition and were technically well laid out. Port-au-Prince itself was a sprawling movie set of a disaster. The twisting, crumbling streets, weaving their way amid snaggle-toothed buildings—bright multicolored dilapidations, mostly two-story French colonial, but the stroller was occasionally brought up short by the culture shock of large official buildings, blobs of concrete plopped randomly. Concrete, the universal symbol of that very comfortable disease, progress.

Poverty was an accepted way of life. The visitor's lasting impression was of bright smiles above bright, shoddy clothes and the perpetual request, in Creole, *"Blanc, j'ai gran' gout . . . donne moi un disco."* They were delighted when given something, and not much less cheerful when denied. Despite their poverty, they were handsome, dignified people whose verve for life was exceptional given the heartrending circumstances. The players stopped walking around town. It was too depressing.

Sunday, there was a masquerade party at the Oloffson, and Jesse, Jaguar, and the boys whipped up costumes. The frenzied dancing in the streets with such sustained, dynamic energies stunned the *"blancs"* for the sheer staying power of it all. The floats were lovely, garish, touching, comical, and occasionally quite ingenious. When the boys returned to the hotel, riding on the tops of several automobiles, waving like movie stars, they were exhausted from the motion and noise levels, which sometimes approached the threshold of pain. After a few of Caesar-the-bartender's superb rum punches, the party, though low-key, made a rather amusing Looney Tunes cartoon, topped off by Wick, who arrived, glass in hand, dressed only in a jock strap, with Christmas lights draped all over him.

Wicko waved to everyone, walked over to a table, sat down, and plugged himself into the wall. He was beautiful. One string blinked "WKO."

The next day, the resiliency of youth erased all traces of the party. Caesar could only shake his head when he saw how they had snapped back. They went right through the locals like a dose of salts, but the Haitians enjoyed the opportunity to test themselves, and their fans howled and clamored for them as much as if they had won. Just the corporeal presence of their countrymen on the court making a good showing against the foreigners was pleasure enough.

Wick had drawn Buster in the first round and was bequeathed a walk-over, which saved him plenty of embarrassment. Buster alone hadn't snapped back—woke up still loaded, drank a ginger ale, and was totally gone again. He managed to make it to the Royal Haitian Club baby pool, put his head on a cushion at the side, and slept in it under a small waterfall. He looked like a koala bear on the "Wonderful World of Disney."

Jesse upset a seeded player in the second round; it was a breakthrough for him and an uplift for the other young players. When Jaguar knocked off an older American player, who had seen his best days and wasn't giving a hundred per cent anyway, the stage was set for a replay of the Junior Wimbledon battle between Jesse and Jaguar. They played a midday match in the hottest sun Jaguar had ever experienced to a near-empty gallery. The few people who did see it were treated to a high-quality spectacle that was close to a grudge match—certainly the points were all begrudged. The boys played their hearts out for three hours. On the changeover at 14-13 in the third set, Jaguar whispered, "I'm dragging my pieces. . . . I'm just all in . . . the heat. I've had it, Jess. You win. Can't . . . run another step."

"Don't talk rubbish, Jaguar. Have a Coke and grab your racquet. I'm not winning like that. Someone told me you never quit. I'm fucking knackered meself. Get out here so I can beat you."

Where it came from Jaguar would never know, but a second or perhaps a third wave of energy supported him through three more games, and he beat Jesse 16-14. It was a good lesson early in life for both of them. Jesse never again would let pride press him into goading a beaten opponent to continue—he would take the luck as it breaks. Jaguar found he had reserves that he didn't know about, or rather, that he was afraid to tap in case they didn't exist. He later remembered his father saying, "You have to reach beyond your grasp. Many times you'll be surprised how far your reach will go." Jag had thought it a load of crap at the time. Jesse had to help Jaguar off the court, supporting him around the waist. They were too tired to notice the loud applause from the now full gallery.

Even massages and whirlpool treatments couldn't put Jaguar back together again in time for his next matches, and he lost both of them dismally the following day. Billy Sherman and Wick had been long out of the running but were hanging around waiting for the others so that they could include them in a plan to take a jeep over the mountains to a place called Jacmel on the Caribbean side of Haiti. They were told that it was cheap and restful—with no tennis. Billy was heading back to England, Wick and Jesse had a commitment in the Virgin Islands, and Jag had been accepted in two tournaments in Florida. It seemed like a good chance to be together in a relaxing atmosphere instead of the back-of-the-mind, razor-sharp presence of competition.

The drive took over five hours of grueling, harrowing determination—hovering on the edges of long, precipitous cliffs, winding down stretches that were little more than goat trails, creeping along riverbeds, careful to stay on hard ground. The sucking mud would have swallowed the jeep up to the doors.

"It's jarring my backbone through my skull," Wick said. "I bet I'm three fucking inches shorter when we get there. *If* we get there. Whoops!" They struck a rut and were almost pitched out over a chasm.

"Whew—that was close," Billy fretted.

"Close only counts in horseshoes and hand grenades," Wick said, the staccato bumping of the jeep chopping up his words. "Mariano did it again," Wick continued. "Whoa baby!" The driver swerved to avoid a large rock and uprooted a sapling which flew just inches over their heads. "He didn't even show up for his match. When I saw him later at the Royal Haitian, he told me he was too busy making money. He says to me, 'What do I care if I'm blackballed in Haiti?' Then he roared at his own joke. The prick made nine hundred dollars at the casino"—long pause for bad stretch of road—"but it did have . . . something good came out of it. One of the locals got a chance to play a big name instead of that turd."

Suddenly, from the crest of the next to last ridge, they saw the tiny harbor, a teardrop-shaped sapphire about forty-five minutes of hard slogging ahead. It was inconceivable to them that anything could have been worth that ride, but Jacmel was. The miniature natural harbor was a perfect horseshoe, surrounded by beach and tall coconut trees, royal palms in thick groves. The water was aquamarine to kelly green, to dark green, to several blues—clear, clean, and cool-looking. Two ancient stone wharfs jutted out into the liquid slap of jewelry that was the bay. The hills that rolled up from the shores on two sides were painted in thousands of nuances of green, and three rivers poured fresh, silver mountain water through the ravines and banana fields, through the washings of lovely, naked, proud-breasted girls and old, craggy, saggy-titted black crones who were sucking smoke through delicately balanced corncob pipes.

The town itself was a beautiful joke. From African straw huts at the edge of the village, the houses changed to a uniformly two-tiered, New Orleans style, with wrought iron balustrades on the second-floor balconies and more filigree. They were in different stages of condition from condemned to perfect, but all were turn-of-the-century at least. The town nestled in comfort and ease on a small rise above the center of the horseshoe. The large, white, Spanish-style church commanded the town marketplace, which was all friendly noise and color during the day.

As they arrived and cut the ignition in front of the little hotel, rather optimistically named the Excelsior, they turned to watch the sunset. The fragile membranes of light melted into darkness with disappearing streaks of alabaster. All of them were touched by the remarkable beauty of it, but Jaguar felt so moved that for a moment he breathed the air of a different planet. The light went quickly, and the stars, because of the lack of dust and the low humidity, seemed within arms' reach.

The tiny eight-room hotel was itself out of an old Joseph Conrad movie scripted just for the "boys." Dinner had a mellow, pleasantly relaxed atmosphere among young men who had shared an adventure and now were one in sweet exhaustion. The themes of the conversation rambled but the tenor was constant and gentle. At one point they talked of the money factor and the pitiful prizes of five hundred and six hundred dollars that used to be given, with thousands being paid out behind the back, and tens of thousands in endorsement payments.

". . . and the height of it" said Wick, "was this Italian bloke, head of his LTA and, would you believe, president of the International LTA, slipping Pietranelli about thirty thousand big ones a year not to turn pro. Not to turn *what?*"

The eerie sound of voodoo drums up in the hills brought exotic dreams, then heavy sleep. The chanting went on long into the night, but they were no longer conscious of it.

Jacmel, with its heavy French influence, could easily have been a small town on the west coast of Africa. With native elegance, the women stepped straight-backed through the streets with baskets, live turkeys, machetes, bananas, nothing too large or too small, all balanced on their heads. Occasionally, donkeys jigged through town with a side-saddle passenger, or plodded along with an enormous burden. There were few proper horses and only four cars, so traffic was busy but gentle.

The boys rode small hill-ponies up into the rugged mountains, high above a river that gleamed with sharp sparks of light far below them. After two hours of climbing over dangerous trails of loose rock and knifelike shards of lava, two hours of exhilarating immersion in the cosmos, as it seemed to Jaguar in his love affair with life, they finally ambled down from their Shangri-la oneness with earth and sky and arrived at a network of four fabulously beautiful cataracts that tumbled into basins, one under another. The bottom one was a shimmering, deep pool of pale, ice-lime water. It was the second time in two days that a realization had been more than worth the effort. Jesse said it for all of them when, after jumping over one of the falls into a basin, he yelled out: "Of all the bloody romantic places on earth and I have to be here with you gorillas. Next time I'm coming back with the woman I love." Le Bassin Bleu.

They said their good-byes at the airport in Port-au-Prince, congratulating one another on having survived the rugged trip. When, years later, Wick discovered that there now was a twin-engine plane once a day that took only fifteen minutes for the trip, his remark was appropriately obscene.

"I'm glad to have met you chaps. I mean it. It'll be a real pleasure whipping your tails on the courts wherever. Soon, I hope," Jag said with firm handshakes all around. Billy had already left for England. Last call had been announced for Jag's plane; he hurried off down the ramp.

Jesse called after him, "I'll be along on part of the American tour later on a bit, Jaguar. Probably see you then. Good luck, Swaggie." Jag waved and was quickly gone. And so had begun their circus-like life.

14

Karen McNally had never seen a tennis match before, or, for that matter, even held a racquet in her hand. All of her energies and hopes were directed to becoming Canadian figure skating champion. It was within her grasp with work and luck. Even in Florida, on her first trip to the States and first real vacation in three years, the seventeen-year-old beauty found a rink in Miami to practice on.

If there is a great cosmic blackboard with the lives of everyone on it, some paths never to cross, others on collision courses for good or bad, Karen's and Jesse's boards showed them crossing a dozen times in the next few years.

Jesse Fraser hadn't played the American circuit for three years, not since his debut on the Jamaican tour. The busy Australian series of tournaments and his two years with WCT, playing mostly in

Europe, had kept him away from the States. Now his contract had expired and he was an independent with a verbal agreement with an American promoter, Phil Katz, who hadn't put a bad step forward yet and who inspired great confidence in the people he handled. He reminded one of Donald Dell in former times, whose loyalty to his charges and quick, astute mind made him a solid figure of respect. Jesse trusted Katz in the same way; if Phil said a handshake was sufficient for a deal, it was.

Jesse's day at the David Parks Courts in Hollywood, Florida, began with the routine, the ritual that Jesse performed every playing day of his life: up at six, a mile run, an enormous breakfast, and at ten a gentle warm-up.

Although Jesse wasn't the big-eared country boy from out-back anymore, though his head had finally finished forming and his hair style obliterated the last traces of his ear complex, even so, he wasn't handsome by anyone's yardstick.

Yet Karen was attracted to him, probably by the glamour of his fame and position as first seed in the tournament (she didn't understand "seed," but she knew "number one" all right). And she was definitely attracted by his thighs. She hadn't had much time for boys in her life; discipline and conditioning are highly demanding in skating, and her parents were strict Catholics. Her awareness of boys was just dawning, and the sexual apects of them, although still a bit of a mystery, gave her heart an extra flutter. Quite simply, Jesse's legs turned her on.

Karen watched his match with her father from the second row. Jesse played the whole match to her, occasionally smiling up on the changeovers. Her father was flattered, Karen, dazzled and choked up with wanting. Eduarde Vector, though, was pissed off. As assistant Davis Cup coach and team tennis coach, he was annoyed at Jesse Fraser's inattention to the work at hand. He was a good psychologist and realized instantly the danger of her presence. Jesse had never, never allowed anything to swerve his concentration, and suddenly here he was, flashing his eyes and swinging his ass like a peacock.

In this, an ordinary quarter-finals of a non-cup match, Vector wasn't allowed to speak to Jesse. He sat there with an impassive face, devising schemes to get this new, unique threat out of Jesse's life.

Somehow, Jesse won. He hurried off the court to catch Karen. She was purposely dallying in the crowd, hoping they could make a magic meeting. She prayed, "Oh God, help him to find me—I'll be good and pure with him if you help." Jesse caught up to them. Surrounded by so many Americans, they took solace in each other's company. After Jesse had attempted what he hoped seemed like a casual "Hi there" (which sounded to him like a confession of love), McNally politely invited him for a beer. They were standing there just staring at each other. McNally felt compelled to say something. After a two-beat silence, a communications breakdown, Jesse accepted shyly, and with his characteristic ineptness, his "Sounds great!" came over at the wrong volume.

He couldn't manage fifteen words the entire time they were together. To make matters worse, Jesse even forgot to arrange another meeting, get a number, give an itinerary—anything. Karen, on her part, was totally infatuated. In some strange way, his homely face glowed with something inner and beautiful.

Karen made her father cancel fishing and come back twice for the remaining matches. She broke her own training schedule, as she was to do often in the future. She and Jesse had two chaperoned dinners together.

After what Wick would have described as a "dirty great row" with her parents, Karen drove to Howard Park in West Palm Beach to be with Jesse, cheer him on, cheer herself up. Oddly enough, Jesse played like a tiger—every shot was a gift to her. He was devastating. Vector's worries were to begin only when they were apart. A couple in love are, to the casual onlooker, sticky, silly, cloying, syrupy. But to them it is an illness, a gnawing, aching illness from which they are terrified of recovering. Mooning and star-treking, constantly telephoning, writing letters, cabling tennis results and love—Jess was a mess. He was, as the English say, "frightfully wet." His tennis suffered so badly that Vector was forced to drop him from Davis Cup consideration for the year. He was twenty down the list in commercial union points and wouldn't qualify for the play-offs. His concentration was so leaky that ham-and-eggers were knocking him off in the early rounds, left, right, and center —this, for someone who had burned up the circuit the year before, winning the Grand Masters' and Wimbledon, was brutal. Few people knew the truth. Most thought that he had been just a flash, a short bullet of quicksilver that was spent. It was love, still in its thunderbolt stage, the first phase of the rocket. Vector threw up his hands and concentrated on Wick, Paul Bell, and Freddie Moore. Australia lost in an early round of the Davis Cup to—Pakistan.

After almost two years of separation, the couple met again in Toronto, where Jesse was playing a leg of the tour with the Orange Group. He was back with WCT again, but well below form.

Although Karen's pursuit of the number one spot in Canadian skating hadn't been as sidetracked by her emotions as Jesse, she at nineteen hadn't realized anything like her ambitions and was struggling along in third place. She didn't quite have that extra edge that made the minute difference—a fact that she would never, in the future, come to face honestly.

At the Sutton Place Hotel at nine on a Tuesday evening, Jesse asked her to marry him. They sat in the lobby in parkas and jeans after a lost match.

"I must have an awful nerve—probably I'm being ridiculous. What else can I do? I have to ask."

"What about my skating?"

"Let's ask questions in their right order. Do you love me?"

"I think so, but what do I know? I'm only nineteen."

"I don't really believe in that puppy-love rubbish anymore. It's the real thing when it's new. Perhaps sometimes it doesn't work when people are very young . . . for many reasons. But the love is real. Do you love me?"

"Yes. Yes, I'm sure I do." She said this slowly, looking into his eyes and feeling reassured at what she saw. "And my skating?"

"If you can think of a way to work it, I'm all for it. I mean you're going on with it. I was sure . . . I always knew that it would be one of the big blocks. But I never had the imagination or con to work out an answer. If you have one, it would be . . . " He brightened suddenly at the realization that he had missed something. She was considering it—seriously considering his proposal. He touched her cheek gently. "Could we make it?"

"Why don't you quit tennis and come with me on my tour?" It was said facetiously, and neither of them entertained the idea seriously.

"You could . . . we could honeymoon in some great cities on my tour, Karen. During the summer months, when you aren't skating. Then—help me here. I need your help. I need you."

"What about this, then? Suppose I say yes, but I try to make one last effort to be champion—a year, say. Then I'll be an old lady of twenty."

"I just couldn't wait. I know I couldn't. Last year I lost Wimbledon in the first round—only the second time in tennis history a defending champion was disgraced like that. I can't keep my head together without you. I want to win Wimbledon again and wipe that out. That or quit. It all depends on you. I love you so much."

"Jesse, you said 'I' about twenty times, then. What about me? Me, me, me?"

"You're right! What do *you* want? Shall I quit tennis?" She knew he meant it, what it meant to him, but neither of them fathomed the full implications. With love, but more than a little pity, she agreed to marry him the next summer, kissed the happiest person in the hotel and maybe Toronto, and hurried out to tell her waiting father, who was sitting patiently in the car.

Jesse telephoned McNally the next morning. He blurted out some qualifications, his good financial situation and total love of Karen. When McNally avowed to be thrilled and delighted, not very convincingly, Jesse's proposal back-up took a strange turn. He swept into a confession. He stammered, "My education isn't all that sound and some bad things happened in my life, in my family, Mr. Mac. My grandfather and some other blokes killed a pearl diver and his wife and stole their stuff. Wonderful pearls. I saw them myself. Maybe you remember a case . . . oh, probably it wouldn't be in a Canadian paper. Eduarde, my coach? Well, one night my father climbed in a window to steal the pearls or the money they fetched. Vector shot him dead. That pearl money helped me to get started in tennis, in life, really. So, you see . . . I don't know what you think about all that, but I wanted you to know first. I don't want any dark secrets."

It was true that Jesse was often haunted by the past horror— the newspapers, the trial—memories, shadows glimpsed, swirling vaguely through mists in the caverns of his mind. Jesse wanted the light of truth and openness to illuminate the ugly facts, freeze them in a kind of bas-relief and not roll through his mind frightening him anymore. It was never completely effective.

McNally didn't know what to say—so he fumbled for words and fragmented phrases like, "Don't worry, son. The love is all that counts." The moment was singularly awkward and not satisfactory to either of them. But at least it got done. In the years to come, McNally always tried to be a friend and good father to both Karen and Jesse. But he kept thinking, "The sins of the father . . ."

15

At twenty-four, Jaguar was still the victim of his own hot loins. Whatever spare time he had—whether on planes, waiting for matches, open evenings, or the empty moments in hotel rooms—was spent in pursuit-and-destroy tactics. The odd-looking, delicious wonder between a girl's legs was, for him, the bottom line of the social contract.

At Hilton Head Island, South Carolina, Jaguar had won a tiny bundle, second money in a round robin with three other men and four women. It was run on a point system; singles, men's doubles, women's doubles, and mixed. Consuelo Alvarro won the big purse, winning all of her events—a nice plus for the women's argument. Jaguar wasn't convinced no matter what. As he said to Al Wick, "She and I won the mixed together. If we'd lost it, I would have won top money 'cause we would have had the same amount of losses but I lost fewer sets. It was a right dilemma, I can tell you."

"You're very noble, matey. She's just a walking furburger, as far as I'm concerned."

"Ah, well, I wouldn't say that exactly. I rather like her, actually. Asked her out. Said she was busy."

"That greasy bloke with the moustache has her all locked up."

"I don't think so, Wick. Looks to me like they're just friends."

"Wonderful. That ought to make you feel nice and worse. Where do you go from here?"

"Up to Raleigh for an exhibition match and a tennis clinic I'm committed to. Bloody nuisance, really. Never can tell, though. Something good might come of it. I'm taking a train. That should be a novelty number."

"I've never been on an American train. I hear they're bloody awful. You had a miserable time with Mariano, didn't you? Christ! He is a swine. If he had to have a rectum transplant, it would reject him. What conniving to win a match."

"Never mind, Wicko. He can't con me. I didn't come here on a hay wagon. Are you leaving tonight?"

"Nah. I'm sticking around for a couple of days. Try to give a little rub of the relic."

"The what?"

"That's what Irish priests call the odd bit of tail."

"Do they really?" Jaguar laughed heartily.

"Yah. Then I'm going up to the tournament in Toronto with my Orange Group. I'm in with a chance, too. Jesse's playing bloody awful. The silly bugger's gone and fallen in love with a Canadian kid. Round the bend he is."

"Here comes my car. All the best, Wick."

"Thanks, mate. You too. Cheery bye, you dirty great Pommie."

Jaguar drove across the lovely Carolina island, finally crossing the state line into Georgia. Hilton Head had made a very favorable impression on Jag. He thought more or less seriously of settling there one day. Very beautiful. So calm.

The Savannah platform was nearly deserted. Only seven or eight people straggled about. Jag boarded his car with that hot feeling of anticipated adventure. His ESP was very seldom wrong about it.

He swung along the aisle, relaxed as one resigned to a pleasant fate. When they saw each other in the dining car, it was all settled on eye contact. She was several years older, perhaps thirty, but her blond-streaked hair set off a perfect face. Her body was lithe, small-boned, and proportionate. Despite the fact that there was another woman at her table, an older lady of indeterminable status, her neutral face crowned by a hairdo best described as menopause fluff, Jaguar ignored the glaring fact of all the other unoccupied tables and sat down.

His English accent charmed and subdued the old gal; his eyes and sensuous mouth did their own job on Mrs. Hot Pants.

"I'm not going to intrude for long . . . just would like to invite . . ." he asked her name by a nod in her direction.

"Charlotte—Charlotte Adamson."

"I'm Jaguar Gray." With characteristic lack of tact, he dismissed the presence of the old lady almost entirely. She wasn't insulted. She was neutral.

"A drink later in the bar?"

"I think that would be very nice. Would half an hour be all right? I have to go back to my compartment."

"Great," Jag said, and patting the old lady's shoulder, he left for the bar.

It was more like an hour before Charlotte reappeared, dazzling him in tight, raw-silk slacks. Her pelvic mound was an impertinent challenge. The motion and metronome of the train cast its spell, and a shared bottle of Chablis—"hock," as Jaguar called it—completed the prelude.

"How far are you going?" she asked.

"Raleigh. I'm playing there tomorrow night. Tennis. That's what I do, you see. Sounds a bit silly, probably, but I'm serious at it. I want to win Wimbledon this year—for my country."

"I don't know much about tennis," she said, lifting a Dunhill lighter to her cigarette. He took it from her and snapped it twice. Charlotte retrieved it and it flamed at first flick. "Don't the English own Wimbledon?"

"Not in the records. Been years. I want it so bad—to win it—that I can honestly taste it. Like I want to taste you."

"We only have a few hours before Raleigh."

"Should be enough. Where are you going?" Jaguar dropped the

hand he was holding onto his thigh and guided it up his leg a few inches.

"I'm going back to Washington. That's where I live—if you could call it that."

"Nice place, Washington. So I've heard." He was paying no attention to what they were saying.

"Washington is a Venus's-flytrap for nice people. Compromise or die. They always get you. They did us."

At that moment a lover's quarrel among three queens flared up at the other end of the car. Jag and Charlotte swiveled around at the piercing exchange. One of them was looking down his shoulder at another, saying, "Listen, you wishy-washy, mush-mouthed, Mississippi faggot . . ."

"I know one little number who's going to get her face slapped," interrupted the second.

"I shouldn't bother—your hand would probably stick," the third chimed in. A glass of something was thrown in somebody's face. The wet somebody flounced out, and it became quiet again.

"Let's get out of here. They're as bent as horseshoes." Jag was angry. The mood was shattered like peanut brittle.

"Would you like to smoke a joint?" she whispered, in sudden serious conspiracy.

"I tried it once and nothing happened. Only made me very sleepy."

"This won't. It's pure Colombian. Come on, where's your compartment?"

"Well, it's little, but it's down this next car." The couchette was small, almost too small for both of them at the same time. While she rolled the cigarette, Jaguar made himself comfortable. He took all his clothes off. She glanced at him with pleasure very apparent.

"You, Mr. Jaguar, are eminently relishable." She was already high from the wine and a pre-dinner toke, and the words were slightly slurred. After the ceremony of passing the joint back and forth, with Jag coughing and gagging through most of it, and burning his lip on the roach, Charlotte said, "Let's go to mine. It's too crowded in here." Jaguar wrapped a towel around himself and they walked through to the next car. Midway down the aisle, she stopped and opened the door to a good-sized compartment. If the car hadn't been shaking and rocking anyway, it would have been from the barrage of love-making. His orgasm had the highest intensity of pleasure he'd ever known. Charlotte was loud about her feelings, too. Sometime during the night the train stopped for a bit. Jaguar looked under the shade and saw nothing. He was stoned. They joined together for another session of When Worlds Collide.

The next time the train stopped it was noisier and brighter outside the window. They sat up together looking about like startled hens.

"Washington! Washington!" A voice floating through the aisle outside announced.

"Jesus Christ!" she said. "You have to get out of here! Get up! My husband's meeting the train."

Which was reality, the dream he had just been catapulted from or the scene around him, Jaguar couldn't say. His confusion turned her fear to anger. "Get out! Please! Now!"

Jaguar swaddled himself in his towel and, forgoing a good-bye kiss, stepped out into heavy traffic in the aisle. Immediately a suitcase swung into his knee. A sweating man muttered past him, pulling the second case at an awkward angle from behind. The second bag got him in the hip. A porter was trying to pass Jag from the other direction. Jaguar joined the stream in the tight right-hand lane, making his way to his car. He opened the two connecting doors and stepped into—the mail car. Halfway through the car, a mail guard stopped him. Jaguar tightened up on the ends of his towel. "I'm trying to find my car. It must be up ahead."

"The engines are the only things ahead. The other cars were taken off at Raleigh."

Jaguar's confusion wasn't immediate this time but settled over him like a silk parachute in slow motion.

"You have to get out of here, buddy."

"You're right about that. But how?" He ran through the car and back to Charlotte's compartment. Trying not to panic, he knocked gingerly but not hard on the door. "I say, Charlotte. My clothes are gone."

"You get out of here. Just fuck off! Now!" She was close to hysteria. Jaguar shrugged his shoulders and stepped into the lavatory. A porter was in there straightening up. A pair of shoes showed under the door of the thunder-box. After Jag explained his predicament, the porter took several minutes to control his laughter, trying to speak through it.

"Well now, don't that beat all. Do you know the temperature out there? Must be around forty-five degrees."

"May I borrow a blanket?"

"Can't let you have one of those. They're all counted and I'm responsible for them. Best I can do is give you these paper shower slippers."

"Fantastic. What's the form next?"

"Yo'all better go in there to Travelers' Aid. I can't do nothin'. We're pullin' out for New York in ten minutes."

"Right!" Jaguar put on the slippers. "Sorry about the tip." Resolutely, he climbed down the steep stairs into the early spring morning. The stares and sharp, cold air put an extra bounce into his step and he entered the enormous station at a half-jog. Washington commuters are not used to near-naked people tortting around in their stations. He amassed a crowd of gogglers around the Travelers' Aid desk.

Cleaning up the story as best he could, he explained his plight to the little old lady with tight blue curls. Her astonishment turned quickly to social censure, but she decided to help him in spite of her disapproval. Jaguar thought that maybe he shouldn't have told her the truth.

Unclaimed clothes from lost and found were brought by a small, skinny black man, who burst into laughter every time he looked at Jag's white gooseflesh. Jaguar was holding his towel and running in place to keep warm. He was very glad to get the bundle, feeling warmer just to have it under his arm on the way to the "Gentlemen's." With a nice touch of *lèse-majesté,* the lost-and-found clerk had included a slightly damaged umbrella. The Travelers' Aid arranged a seat southbound to Raleigh, and also located his clothes for him. Jaguar tried to give the woman a kiss for her kindness, but she turned away in horror. As he went back to the platform, swinging the broken brolley, he looked as though he belonged on top of the train rather than in it.

That night in Raleigh, he rocketed from the ridiculous to the sublime, winning his exhibition match in straight sets.

16

On the train, Missy Teaford tried, by studied control, to get a firm grip on her nerves and examine herself, her position, calmly and logically. But she was scared. Afraid that she would do something mad. Perhaps even throw herself from the train. The man across the aisle was looking with keen suspicion at her. He knew that within the exterior of cool, all-American girl, within the wrapper of aloofness and style in a twenty-six-year-old body, she was confused, directionless, and scared. They probably all knew.

She looked out the window. Just above the sharp, snowy peaks, the pale scimitar of crescent moon was barely visible against the milky blue of the late afternoon sky. Waiting for its turn. Everything in life, waiting. The snow cascaded down the side of the mountains like great bolts of pure satin, highlighted by patches of brilliant sunlight, glossing the harshness of the crags and hiding fault-ridden ravines very much in the way that illusions obscure the brutal ugliness of life. The trembling power of the Alps was even exaggerated by their very impersonal indifference—a godly trait.

The train sped through a tunnel in the arm of a mountain, only to explode from the darkness into the emblazoned white vastness. The grandeur and chiaroscuro effect made the fantasy of the mountains more pronounced and her sanity more precarious. She had written on the back of her ticket folder, "With nothing to wait for and without the 'concrete realities' always appended to the 'illusory goal,' there is nothing but the weightless falling through time." Her racquets were hung up, perhaps forever. They were locked up and she was released. All the artificial crises and pressures were locked in there with them, along with the nebulous, rather ridiculous goals.

Doctor Morgenthau said that after some time, when the mental lesions had knit and her approach to the game was fresh and in healthy perspective, she might even return to big-time tennis, probably with a sounder game.

There was a good powder on the slopes when she arrived, and the crowd seemed very attractive. Once she'd installed herself in the bungalow and changed into her ski clothes, she immediately lost the uncertain feeling of the encroaching newcomer, and had there been another train arrival that evening she would have greeted it with the good-natured tolerance of a native and veteran.

It was very surprising to her that after the terrifying (for her) trip across the mountains, knowing what a delicate membrane she had maintained between control and hysteria to hold back the flood of panic, she could step out of that strained personality like a pair of pants. In her ski things, clinging tightly to her like a new skin, she crunched up the softly moon-tinted road toward the lodge. Without trying to be logical, she felt simply, not that her old self had been reconstituted, but that it had been discarded and replaced by a new one. She was not the sum total of her experiences or memories; she was born the moment she'd stepped out of the bathtub.

A sleighload of bundlers jingled by with a well-wishing roar of good will. She answered with a wave and a spontaneous giggle, sharing their fun for an instant. Missy was so happy that she swelled up and, with an exuberant rush of air, burst into song, stopping only when she reached the steps of the lodge a half mile away.

Although the main room of the lodge was sparsely lighted and a fire was going, it was still dimmer than outdoors, and it took her a few moments to adjust. A form bounded out of an easy chair and took her arm. "Why, you poor, cold, lost little thing! For God's sake, come over here close to the fire and let me get you something to take the death chill out of you if it isn't too late."

He propelled Missy to his chair, scrunched her down into it by the shoulders, and leaped off to the bar. In the glimmer of the firelight she saw the good-humored smiles of several young couples who had either not gone upstairs to change yet and were a little advanced into the booze, or others who had changed already and had come down for an early cocktail.

"Frankie is allegedly harmless," one of them volunteered. "I'm Richard Bart, and this is my wife, Anne. We'll protect you if he comes on too strong. He gave Anne here the rogue-elephant technique the first day we got here. She was gasping for breath when I arrived and pulled the plug on him."

"How do you do." Her smile was weak. "Jesus Christ! What a nut case," she whispered with a slight tremble of voice. "My name is Missy Teaford. Don't let him do that again, would you please?"

"I don't believe it! It really *is* Missy Teaford! We've seen you play a hundred times on the box, and Bart's seen you play in person at Forest Hills."

"Well, my God, yes! But I didn't expect ever to meet you in person. My seat was so high up, the court looked like a postcard," he said, peering at her closely. One of the wonderful squares who make up our wonderful country, she thought. Solid and unsophisticated.

"I told Anne how great you played and how you should have won the finals and all," he went on, ". . . and we went to see you when the Virginia Slims came to Indiana—but you'd scratched."

"Yes, I was quite ill." Oddly enough, this reference to her playing, to tennis, was painless. Her family and the doctors had tried to discuss it normally as a natural subject, but she considered that for what it was, a tense attempt at therapy. This was the first time a stranger had broached it.

"We looked through *World Tennis* magazine for a listing of you on the circuit, but you'd dropped from sight," Anne said. "Bart's a big fan of yours and I'm a little jealous. I'm only starting tennis and I'm uncoordinated as hell."

"I'm sure you'll get on to it if you get good instruction at first."

By this time the group had gathered around, ready to lionize their private celebrity.

"What happened to you?" someone asked bluntly.

The firelight danced on the shadows of her face. Her hands twined and untwined in her lap, and her voice started out weakly at first.

"I was forced to retire because of my health—the strain of traveling and . . . and other things . . ." She trailed off.

"It sounds like an Ingrid Bergman movie," one of the pleasantly amorphic faces contributed. It brought light laughter, but Missy's face was white and taut, and they tapered off into an uncomfortable silence.

"Sorry to be tardy with the toddy, but I got sidetracked at the bar," Frank blurted, his smile all teeth. He had just missed being handsome, but this nuance was unknown to him and he took his desirability on blind faith. He handed a steaming, musky-scented glass to Missy and sat down on the arm of the chair, placing a familiar hand on her shoulder. She leaned forward and tried to sip her drink but it was too hot. The evasion of his advance didn't go unnoticed.

There was a lightness once more in the group as they talked about skiing now and in the past years. Whenever possible, Frank talked about himself. The time before dinner passed brightly and pleasantly —until Angie Redfield bustled in out of the snow. Missy recoiled. Her heart beat heavily and painfully. Angie didn't see her at first but clucked rapidly and badly in German to a blond, Bavarian-looking man she had in tow.

"Missy! It's Missy Teaford. My God, what in the ever lovin' blue-eyed world are *you* doing here?" she gushed when she finally spotted her.

Angie was a former junior champion who had recently joined the tour. Missy had followed her rocket closely through her first year, before the doctor had cut out all exposure to tennis and the other causes of her illness. Angie sat down on the edge of the chair, still warm from the recently departed Frank, who had gone back to the bar. "Where have you been, stranger? I sort of heard you'd been sick and weren't playing anymore." Her elaborate Eastern-college accent underscored the tactlessness of the remark and she hastened to cover, "but you look simply marvelous now! Just marvy!"

Without bothering to introduce the man with her, she made a great to-do about inviting Missy to join them for dinner.

"I'm not really dressed for the evening. I put these clothes on for a kind of shakedown run. Maybe we can do it another time. I'm going to be here for about ten days."

"Well, tomorrow night for sure!" Angie was adamant about it.

"All right then, tiger," Missy chuckled, "I'll met you here tomorrow night about the same time, give or take five minutes." She drained off the last of her buttered rum, and after a stiff handshake from the young man, left them for the stingingly refreshing briskness of the Alpine night.

Without being too intense about it (she was *not* to be intense), she wondered about the wisdom of being with anyone here when she was supposed to be resting and recovering, especially someone from the tennis world. Having weighed the merits of several possibilities, she decided that this might be the gentlest way to ease into normal life without pampering herself too much. The bravest way to initiate her "vacation." In any case, she was committed. She smiled and hummed to herself while changing for bed. She even laughed out loud as she remembered Frank's apology for not being free the next night, accompanied by a promise to make it up to her. His was an ego so ironclad that it was almost enviable for its smug security. He was that rare beast, the amusing bore.

The paneled dining room was a little smoky from the choked-up fireplace, but the well-cured logs were sweetly pungent and the soft haze made the room seem more intimate. The wooden tables shone deeply from the polish of years of human warmth. Several of the people she had met or seen already were there, including Frank, who gave her a gratuitous wink over the shoulder of his date. "What an asshole," she thought, and barely nodded.

Angie Redfield and the big blond waved from a table in the corner. The Bavarian leaped up as Missy approached and, with a flourish of stiff manners, held her chair and snapped his fingers for the headwaiter at the same time. He ordered a chilled bottle of Riesling, tasted it, and then, incomprehensibly, lapsed into attentive silence for the rest of the meal.

Angie was sweet, gregarious, and really good fun. Her stories were spiced with rich, new collegiate slang and hip argot, and she had the native timing of a professional comic as she told them. Her response to Missy's anecdotes was spontaneous, and she laughed easily and warmly and the meal passed quickly. The wine was tingling through Missy's whole body and she felt a flash of giddiness, tinged with a touch of sadness that this wonderful moment, like everything else, would pass. The Bavarian, who had been quietly absorbing the pleasure of the two, excused himself after the fifth stein of beer.

"You know, you haven't told me what you're doing here right at the height of the tour, Angie." Missy said this while peering with one eye through her wine glass, held up to the candle.

"Oh. Right on. There was an exhibition over in Gstaad. It finished last week and I had six days before the next tournament. In Hartford. We all drew straws for who was to play it and I was one of the losers. I hate snow. At least I thought I did. Really, it was all just great and the people there and here are terrific. You know, but really out of sight." Angie glanced over her shoulder. "Now, for Christ's sake, don't leave me alone with that wonk, Missy. He's a perfect gentleman when he's sober in the light of day, but these ski instructors turn into vampires at night."

"All right, I promise. He's not such a bad-looking guy, though."

"I have absolutely no eyes for him. He's like a wind-up doll in everything he does. He probably fucks like one. As I said, on the slopes he's fine—painless at dinner. But then—curtain! These guys have every American girl who comes here marked as an easy lay. The idea that there might possibly be a little decency among our female output just never occurs to them."

"Plenty of girls, maybe most of them, do have getting laid at least in the back of their minds when they go on vacation. Now you *know* that's true."

"I don't agree with you. Anyway, the last one told his friends 'Zat

I was probably und lesbiand.' That's always their face-saving rationalization. Did you come here to get screwed?"

Miss squirmed in her chair. "No," she said, softly, "I certainly did not."

"Well, I frigging well didn't either. That's two out of two. So good for them and hooray for their egos. Cool it, here he comes."

He reached between his legs, pulled his chair under him, and smiled brightly. Whether at them or at the stein he was picking up they couldn't be sure.

He really is very nice, Missy thought. Good-looking and very patient, considering he had been excluded from the conversation all evening.

Angie signaled to Missy with her head and stood up. She smiled at Anton (it turned out that he did have a name actually), and in fractured German told him that he was her guest that night for dinner and that she would see him the next day, no doubt. She was so charming in her halting attempt at the language that it was impossible for him to be angry. He stood up, not as steadily as he might have, and snap-bowed his good nights to them.

The sting of the snow-scented air merged with the heady wine, inducing a short spurt of rhapsody. Angie took Missy's arm as they trudged through the snow. They chattered away delightedly, as if they'd been school pals for years. Although it was still early, Missy was tired and very reluctantly accepted Angie's invitation for a hot schnapps nightcap.

"Missy-face, you can't imagine what it's like trying to make conversation with those cozily married ingrown toenails here. Like they're vegetables. Really! I'd rather struggle along in German. I was so glad to see you I could have cried."

"I just really can't stay for more than fifteen minutes, Angie. My eyes ache from wanting to close. It's a fight to keep them open."

Angie brought the drinks into the sitting room of her cottage. She put on a Bavarian folk dance and whirled around the room to it, seeming not to need a partner; only an audience. She was an excellent dancer and her steps were delicate and smooth as she performed the intricate movements. In the middle of the dance, she ran to a closet, pulled out a fur cap and, returning with a skip, plopped it down on Missy's head and pulled her up by the elbows. The basic step wasn't that difficult, but they made a ridiculous-looking duo; like Batman and Robin. Missy, particularly enhanced by the cap, towered over her petite partner. She made the appropriate Teutonic leers and they both laughed until the tears rolled. Suddenly, Angie stopped short and, pulling Missy with unexpected strength, kissed her full on the lips, thrusting her tongue deep into her mouth and digging her nails into Missy's ass. After a short struggle, Missy pried her arms away and pushed her back. The music had stopped. They stood facing each other for a temple-pounding eternity. Missy's emotion was less surprise or disgust than it was disappointment. Angie's feelings fluctuated between frustration and revulsion with herself. Finally, she broke the silence.

"I hoped I wouldn't do that. Not tonight! I promised myself I wouldn't . . . that I'd try not to. . . . It's all so sickening, isn't it?"

Missy sat down and picked up her drink without replying. Angie lighted a cigarette with trembling hands and went on: "I hate the concept of it and I can't picture myself in it at all. I'm so feminine, really I am. In every way! But then there . . . there the hell I am! I went to this analyst in the city, New York, and I actually don't know that the creep believed me. Why I was there and all. He gave me this psychological garble-de-gook about misdirected friendship which all melted down to the fact that I was a queer. I hate it!" She wasn't sobbing of sniffling through all of this, but tears streamed freely down her cheeks. "I'm quitting tennis too, I guess. Maybe not. It's my thing. For Christ's sake, I don't know what I'm going to do!" Angie said this vacantly. Not to anyone. Perhaps not even to herself. When Missy put her hand on her shoulder for a moment before putting on her parka and leaving, she didn't even notice.

The morning's skiing had been superb. The snow was perfect; the scenery as awesome, as august as Missy had imagined it to be. As to her form, she hadn't broken anything yet, and that alone was gorgeous.

"Good morning," Anton shouted, jogging to catch up to her. "Are you having lunch now?"

Missy was amazed. After the previous night's silence at dinner, she didn't expect him to be speaking English, and here was this pure Oxford accent with only a tiny taint of German. He caught up to her, a little out of breath.

"Hello, hello! Ah . . . phew! You have very long legs. I thought that was you back there and finally had to run to overtake you. Now then—whew, I am quite out of vind. You must allow me to take you to lunch. I was very naughty last night. Here, let me take your skis."

"And where did you learn your English, Mr. . . . ."

"Anton. You must call me Anton. Yes. I went to several good schools in England. Ending up ignominiously in hotel management. My father owns the lodge and the restaurant where we dined last night. I spent a great deal of time in Jamaica as well, though I suspect that it didn't help my accent very much. About last night. I *am* sorry. Of course we all know about Miss Redfield—Angie. I didn't know if you were one, you know, like her, or not. I wasn't exactly spying, you see. I really don't understand much she says in German and she spoke in terms with you I couldn't quite get. Also, I didn't want to interrupt. You understand I feel I have a responsibility. To my father—to the reputation of the place. I must admit that she proposed a bit of a challenge as well."

His smooth and more than somewhat insensitive delivery rankled Missy, and her reply was as icy as the ground they trod.

"And what am I, Mr. Anton? A challenge or a responsibility? I mean, have you made up your mind whether I'm straight or not?"

"Anton, not Mr. Anton. Ah, yes. I knew after a while that you were a true woman and that I desired you very much. Would I be less than honest if I pretended that I didn't want to bed you?"

"*Bed* me? You schmuck! You want to *what* me? Angie's right about one thing. You bums are all alike."

"My dear Miss Teaford. I take issue with the term 'bum' which in *good* English means the part of the anatomy which makes the *sitzmark*, and I take special exception to the fact that someone of your grace and sophistication should have the bad taste to confuse simple good manners with an inelegant and shamefully unreluctant, what is the word, proposition. I asked you to lunch. What do you say to that?"

"Fuck off, buddy!" She grabbed her skis from him and strode down the hill toward her cottage. Her flush was much too full to be merely the result of one emotional stimulus. It was the combination of flattery, distaste, and—one of the major parts of her illness—lust. Although the doctor had convinced her that she wasn't a nymphomaniac (he resented the term), he only dampened a constant inclination to horniness which she felt. He did instill a stronger fiber of selectivity and discretion, though. To cushion what he called (it sounded silly coming from the prissy Dr. Morgenthau) her well-developed sense of carnal intensity. "How's that for an intellectual soft-shoe?" she had thought at the time. "Wow!"

In her cottage, feeling absolutely nothing, she sat down to write a note to Anton, feeling that she had been undoubtedly too abrupt. After all, he was trying to be nice and it wasn't his fault he was an impercipient clod. She wrote on a piece of lodge stationery:

> Dear Anton, whatever,
> I'm sorry I was so short with you. With a chance to weigh the insolence and incidence of your remarks, I might say that, all in all, it was much less than a velvet presentation of a very silly and cumbersome idea.

"Yuk! I never could write letters." She crumpled the note and threw it into the wastebasket.

The incident had drained her enthusiasm, making further skiing that afternoon out of the question for her. After a purifying shower and change of clothes, she went to the lodge bar for a "pick-me-up." With a little giggle, she thought, I mustn't even use terms like that; what would Doc Morgenthau think?

The place was almost empty except for a couple of non-skiers at the bar and Anne Bart, who had sprained her ankle slightly and was out of combat for a day or two. They were glad to see each other.

"Hi there, Miss Teaford. Come over and cheer me up. I feel like a child with chicken pox watching all the other kids playing outside."

"I'm Missy. Miss Teaford sounds like an elderly school teacher. Sure, I'd love to join you. I'm not feeling so crisp myself."

She pulled up one of the heavy wooden chairs from the next table and plucked a cushion from a chair across the room. "What are we drinking today, Anne?"

"This is a large Manhattan. Ordinarily, I can't even stand the smell of bourbon, but it's passable mixed up like this. Have one. You look like you could do with one." The barman came around to their table. Missy indicated with a nod toward Anne's drink that she'd have the same.

"How's your ankle? Is it broken?"

"No, just a tiny sprain. It's a little humiliating. I did it going to the toilet. Don't ask me how. It was really tricky."

They laughed, then were silent, and Missy sampled the drink the barman set down professionally in front of her. She smiled her appreciation and he walked off with quiet pride to the bar.

The conversation was desultory in the beginning, but Missy could sense the turn the talk was taking as Anne eased into her fourth or fifth Manhattan. "It must've been a heavy blow to you after years of practice and hard work, and, you know, building up your reputation and that sort of thing, well, to get sick and have to give it up. I mean . . ." She left it drift. Missy didn't rise to the bait but busied herself trying to lance the cherry at the bottom of her glass with a straw.

"Did you get anemic or something like that?" Anne pressed. " 'Cause in that case you could always take a rest and build yourself up again."

"No, Anne. So we don't have to dance around the subject too long, I had, in the layman's terminology, a nervous collapse. Breakdown, if you will."

Anne was genuinely chagrined and stuttered through a few false starts. "I . . . please believe me, Missy, I didn't mean to pry . . . I wasn't . . . really, I had no idea. Please forgive me if I've done a bad thing. I had no intention."

"Of course you did, but it's quite all right, Anne. There was no harm done. Curiosity and gossiping were given to women by a fairy of ironic disposition. No. It was probably good for me to say it out loud. The pressures of my parents, of tennis—unrequited love—like too many of them. I fall in love too easily."

She chuckled at that and took a generous belt from her fresh drink. Carelessly, almost jauntily, she moved over to the covered spinet in the corner and stood for several seconds with her hand on it tentatively, as one pets a strange dog. With a single motion, she yanked the cover from it and opened the lid. Then she sat down before it, like a racing driver sitting in his car for the first time after a bad accident. She hadn't played for years since her prodigy days. Her mother and father had made it too tough to continue. The piano was slightly out of tune but too slightly for the average person to tell. Missy launched into "Stella by Starlight" with a concert treatment going from the basic theme into counterpoint, whirling off into long, light, succinct arpeggios, and always maintaining a deep, rich interpretation. When she'd finished, she sat for several moments looking at the piano, a chilling sweat on her forehead. Her sigh came from the depths of her soul; it was the exhalation of an evil—she felt empty—limp, but delivered. If she could do *that* again, she could easily go back to tennis.

The people in the bar (several had drifted in as she played) gave her a small but enthusiastic ovation. Anne was clapping and weeping drunkenly.

"Terrific! But it made me feel so sad, the way you did it, and I don't know why."

Missy stood up and walked back to the table. Anne had gone totally dumb. Frank walked over to their table from the door, where he'd caught the last part of the performance.

"Great, great, Missy. Waiter! Another round over here. Missy, my precious, how about dinner with me tonight. Then we can go to the amateur night they're giving at the lodge. You can play there and maybe win first prize. What do you say?"

She turned slowly in his general direction, and, without looking up at him, blankly shook her head. All of the bluster went out of Frank at the summary rejection. He was furious.

"Well, I can't say I'm surprised. I heard about your date last night. How was it?"

Missy suddenly had a great interest in looking at her drink. "Why don't you just go away. You give me the creeps."

Frank sneered, threw back his head, then, turned on his heel and left.

"Don't pay any attention to that donkey, honey. Could you give me a hand getting up to my room? I'm a little too shaky to make it by myself with this gimpy foot. You'll like the suite. It's real Alpy."

Her limp was accented by the hundred-proof Manhattans and they looked like Uncle Wiggly in Connecticut going up the long, winding stairs together.

"I have a bottle of brandy here someplace, Missy. Just sit down over there and I'll dig it out. It's too late for dinner, I think. I wonder where the hell Bart is."

The brandy revitalized Missy's downer with a tingle. She sat quietly as the tension left her body and throat while Anne, far into her cups, indulged in a monologue—the tough life she'd had and was perpetuating with Richie Bart.

"You know something else," she said, pouring herself another brandy, "today's our wedding anniversary. Do you think for one moment he remembered? Like hell he did. I'll tell you something else. That guy's been a washout ever since I've known him. He washed out of the Air Force cadet school after only two months. Something about psycho-motor coordination. Psycho is right. His father wanted him in their law firm. D'you think for one moment he finished law school? In a pig's ass! Now get this. He went to California to train for the Peace Corps . . . the *Peace Corps*, already! Now here's the big bag. He washed out of the Peace Corps. Isn't that a laugh? His attitude wasn't right. That's where I met him, right? So we went out quite a lot and he got his wild hair about Oriental languages, so off we go to Hong Kong. Now he works for this rinky-dink import-export outfit working for peanuts—and in all that time I'm not even pregnant! A complete washout."

She put her glass down with finality to punctuate her diatribe almost simultaneously with a heavy thump from the next room. They both started at the noise and slowly rose and walked toward the bedroom—Missy a little behind Anne.

There, in the gloom of semi-twilight, stood Richie, his chin on chest, his arms dangling limply at his sides. He was sobbing gently. On the floor beside him was a gaily wrapped package. Anne gave a short cry of anguish and threw her arms around him, kissing the tears on his cheeks and murmuring apologies. As Missy opened the door to leave, she could hear him say through the heavy sobs something that sounded like ". . . out lushing it up with that lesbian . . ." She closed the door quietly behind her and trudged wearily to her cottage.

It *was* too late for dinner and she didn't have the stomach for it anyway. She considered ringing for a sandwich but discarded the idea. After an unrewarding bath, she took two sleeping tablets and climbed into bed, the bottle of Valium still in her hand. She caught herself staring at it; wanting to blot it all out—in instant sleep.

It must have been sometime around midnight. Missy awoke to an incessant pounding on her door. Drowsily, she switched on her reading light and picked up her traveling clock. It had stopped. She pulled a robe on over her nightgown and opened the door a crack.

"Yes, what is it?"

Frank pushed into the room.

"I thought you were going to let me freeze to death out there," he said with a vicious smile. "I've been trying to rouse you for fifteen minutes. I thought something was the matter. Are you going to offer me a drink?"

"I'm going to offer you nothing but my contempt, you disgusting slob, now get the fuck out of my room!"

As he pushed her back toward the bed, his walk, desperately normal, had the studied liquid motion of a drunk trying to appear sober.

She pulled away from him, trying to get to the phone, and as she did the top of her bodice came away in his hand, exposing her firm, honey-colored breasts. His anger changed to real lust and hot dryness caught at his throat. She tore his face with her nails and bit deeply into his ear, but the passion pumped adrenaline into his heart and he felt nothing. With one hand he pushed her on the bed, pulling the phone out of the wall with the other. She rolled off the bed and tried to crawl to the door. He was on her in an instant. He ripped the remainder of her nightie from her as they wrestled on the floor.

"I've got you now, you snotty motherfucker!" he gasped as he clawed two fingers up into her dry membrane.

Her attempts at any kind of control left her and she screamed like a demented creature. Frank clamped his hand over her mouth and rolled himself between her legs.

"You won't be dry long when you get this baby inside you, you stuck-up shit."

Suddenly his weight was lifted from her, and she was half-conscious of a brutal fight back and forth across the room. Anton brought his knee up powerfully into Frank's crotch. He chopped him across the bridge of his nose with the side of his hand. Frank's wild right missed, but his kick caught Anton painfully in the shin and doubled him over with a short grunt. Frank hit him on top of the head, knocking him to the floor on his knees. His kick, aimed at the head, struck Anton obliquely in the chest and he rolled over for an instant. Frank's next kick missed completely and his own impetus carried him right over Anton and into the corner. Anton jumped on him with both feet, then jammed his fingers into his eyes and grabbed at his balls. Frank somehow reached around and got Anton by the hair, and, blindly, bit deep into Anton's throat. Anton yanked back and butted him in the face with his head, chopped him in the Adam's apple, then rolled to one side to get more leverage. Frank still tried to get to his feet. Anton hit him solidly on the side of the head with the poker and Frank lay very still.

Missy had pulled the quilt around herself, and was rubbing the blood from her badly scratched arm with a face cloth.

Anton stood panting and heaving for a full minute before saying, "I don't think he's dead . . . he seems to be breathing . . . I . . ."

"I don't care whether he is or not."

"I'm terribly sorry, Miss Teaford . . . really . . . very sorry. I hope there's no scandal."

"You *would* have to add that. Don't worry. I don't want anyone to know this either." She uttered a short, dry, mirthless laugh.

"I mean that I am sorry it happened to you, too. I didn't mean it to sound . . ."

Frank moaned and moved a little on the floor.

"That's all right. Thank you, I guess . . . really, yes, thank you. Now get him out of here."

Anton took the heavy body and dragged it out the door and pitched Frank head first into the snow. He tried to get to his feet but fell unconscious again. Anton nodded good night. A thousand unsaid, complex things formed on his bruised face; it went blank again and he went off into the night to get a stretcher for Frank. He dragged his own badly beaten body through the snow, limping in pain.

Missy sat on the edge of her bed. Dawn would be, at best, an unwelcome intruder. No! She went to the armoire and took down a dress, her bag, and a pair of boots. She had slipped many notches from the unexpected liberation of that first afternoon. She put the dress and her ski boots on, slipped into the parka, and walked to the medicine cabinet. She looked at herself in the mirror, brushed a swatch of blond hair from her forehead and drained the water glass of Kümmel she'd poured for herself. Missy took the skis from the hall and opened the door.

Outside, the lift was ghostly shadowed, as if it hadn't run at all that day but had been left there from a previous civilization. She stared up the north slope. Masked in a dramatic penumbra, it seemed tactile and inviting. She stood entranced and undecided.

The moonlight around her gave the slopes a pale, shimmering luminosity.

She started a criss-cross climb, the wind knifing through her thin dress, then suddenly sat down squat in the snow. The moon dropped off behind the furthest crest of the range, deepening the dark swirl of madness she trifled with.

"I'm stronger than I thought. I'm too strong for this crap!"

Her decision made, Missy shushed slowly off down the hill to the railway station to wait for the first train. She would send for her things later.

17

The men's first-round singles matches were under way at Queen's Club on a splendid, crisp June afternoon. The shirt-sleeved crowd was a faceless study on a blue-green background. A Degas mood. Every so often, Jesse would walk to the dressing-room window and lean his one hundred and eighty-three pounds on the palms of his hands, checking the score and watching bits of the match in progress on court one. His was second match on. His progress check allowed him to know when to dress. He preferred to be ready before being called over the loudspeaker.

The Italian, Berconi, was winning easily, putting on a sparkling display of genius—touch and power out of a magician's mixed bag. Always unpredictable. Jesse often suspected that Berconi himself never knew what his game plan was. The artistry was conceived of the moment in a madman's mosaic.

Wick clomped into the dressing room wet, grass-stained, and bedraggled, still chanting a soliloquy started somewhere on the stairs: ". . . and there I am down a break in the first set and 1-4 . . ." he continued to speak through the wet shirt he was tugging over his face, ". . . and this Irish twit of an umpire says—no linesmen, ya understand, just this Irish charlie sittin' up there like a praying mantis in a blazer . . ." He plopped down on a bench and went to work on his laces. ". . . he says, 'Play two, please, I couldn't see it.' "

Jesse interrupted without turning away from the window. "Well, you have to admit you do have a thumping serve. Without a linesman? Pretty tough."

"Hear me out! I say to him, 'That's the third mistake you've made so far, mate!' And he says right back to me, 'If you'd only made three mistakes you'd be winning this fucking match.' I dropped my racquet. Good job there weren't many people around that far-end cow pasture I was on 'cause I could see he was embarrassed. It just slipped out. He was right, you know. I ran the set out and broke twice in the second. But can you imagine that from the chair?"

"Not really. That sort of wisdom generally never gets a voice. The Franco Berconi Show is almost finished down there. I'd better get my kit on." ·

Jesse passed Berconi on the stairs, congratulated him, and accepted the half-hearted "Good luck" wished him. Even after victory, Berconi wasn't relaxed. More than simply the elation of winning, his highstrung, nervous aura, the bright eyes and taut nerve endings that seemed nearly transparent were to Jesse the dazzling phosphorescence of decay. Berconi needed a rest and there wasn't any time. His peak has been reached too soon. Too many tournaments—too much money—too much pressure—all with no inner strength or calm to draw on.

Jesse shrugged his shoulders, took a deep breath, and joined his Belgian victim. They entered the court together.

It was an amusing if short match. The Belgian had good strokes and good manners. Jesse had power, experience, and killer instinct. Although born too late to be a student of his, Jesse had had the benefit of the backwash of the great Harry Hopman, probably the greatest tennis coach in history.

On one point, Jesse mis-hit a topspin backhand and whipped it miles into the stand. When the boy let it go, Jesse shouted, "Marcel!

You have to hit those. How do you hope to improve if you don't practice!"

Everyone, including Marcel, roared with laughter mainly because it was uncharacteristic for Jesse to open his mouth on the court.

One and love, Jesse was back upstairs so quickly that Shep, who had been warming up on an outside court with Missy Teaford, was five minutes late. The bullhorn was announcing him again and again, and he was answering unheard from the shower, "I'm coming, goddamnit! I'm coming!"

His match was more difficult but not really testing. Mariano's was equally easy. In fact, all eight seeds ended the day in straight-set wins. Only Jaguar came close to dropping a set after having fallen in love with a girl in the crowd. During one of the changeovers, he furtively asked Billy Sherman, who was standing by the chair, "Who is that dolly over there in the pink sweater?"

"Which girl? I don't see where you mean."

"Watch where this wide forehand goes and find out!"

Pancho O'Brien served wide to Jaguar's forehand. Jag almost decapitated a girl in a pink sweater with a blazing return out. Billy rushed over to see if she was "all right."

Unfortunately for Jag, Billy was a bit oversolicitous and finished up with a date with her himself. Jaguar kept an eye on their budding romance and his increasing agitation clouded his concentration. Pancho was suddenly serving for the first set at 5-4. Jaguar remembered where he was and battled back into the match, finally beating the Chilean 8-6, 6-2. He was severe and unpleasant with himself afterwards for the lapse.

Conseulo Alvarro was upset with him too. She had been a perceptive and unhappy onlooker to Jaguar's interest in the girl in the crowd. No, she hadn't yet made up her mind about the most important move of her life, about marrying Jaguar, but nonsenses like this made it a closely run thing. Her affair with Jaguar had been as closely guarded a secret as her private training sessions at the court of a rich Argentinian friend and patron in Oxfordshire. She loved him, all right, and he her, in his own way. But she wondered whether she could ever accept Jaguar's way.

Without waiting for the usual congratulations to Jag, Consuelo pulled her Stewart-plaid shawl more tightly around her against the sharp wind that arose with the late evening cloud cover. She tossed her glossy black hair back from her almost too aquiline face and pony-stepped toward the tearoom. *Consuelo is fucking angry,* she thought in English.

The dressing room was the normal bedlam of a big gathering of semi-matured athletes. Apparently grown men chasing around trying to snap one another in the ass with wet towels—playing pranks—bragging about sex, earning power, playing ability, social contacts, political inroads.

Wick watched Billy Sherman running about telling anyone who looked faintly interested about his afternoon's success. Not his one-sided first-round win, but the upcoming date he had with the "pink vision." Wick called across the room in an exaggerated Australian accent, "Billy! Stop jumping around like a fart in a frying pan and get over here!"

"What can I do for you, old man? Actually, I don't have a super-abundance of time. You see, I am *frightfully pushed.*"

"You're frightfully shit, mytey. I mean, what is this hopping about like a spare prick at a princess's wedding? Is it your first *assignation* then?" (the last in Wick's concept of BBC English).

"Beautiful girls don't come down the carriageway every ten minutes, Wicko, and this girl's a winner. She's sensational!"

"Leave Billy and his girl alone," Pancho O'Brien interceded.

"*His* girl! His girl indeed! He's never so much as said 'How de do' to her."

Wick continued while shaking talcum on his feet. "Jag told me he introduced you and you took advantage."

"That's wonderful, that is!" Billy said, swinging his racquets back on his shoulder. "Some introduction. He nearly took her bleeding head off. Anyway—I'll see you chaps later. If I return at all. Cheery bye." He paused. Then: "Don't be jealous, Al. You still have yourself," Billy called over his shoulder. "And masturbation is never having to say you're sorry." Sherman was gone.

In a partially serious, offhand manner, Al Wick enticed a towel-wrapped Shep into a discussion on the consequences of having sex before playing sports. He hit on exactly the right person, because Shep had some very definite theories on the subject—most of them with valid premises and invalid conclusions. Schopenhauer once said that most men argue for triumph and not for truth. Shep's conclusions were so foregone that his bad examples might just as well have been skipped. It was his theory, reinforced by dozens of specious references, some blatantly made up on the spot, that man's killer instinct, the will to win, was hurt, or at least subdued by orgasm.

"Like your Teddy Roosevelt," Wick interjected. "Walk stiffly, but carry a big soft."

"That isn't exactly the phrase that would have sprung to mind," Shep snickered with a half-smile below very raised eyebrows. He went on to remark that male opera singers were well known for their bad performances after sex, whereas, sopranos, especially coloraturas, warbled with subtle vibrance, their tensions relaxed and glands released.

"It gets the juices jumping and the hormones humming," Wick sang, clicking his fingers in a clumsy flamenco.

"Wicko, you're one of a kind, I must admit." Shep grabbed his bag and high-stepped toward the door.

Now Al Wick turned his attention to Jesse and said in dialect, "Yah gonna sink a fyew tubes with the mytes, Jesso?"

Jesse continued brushing his lusterless hair and answered in kind, "No, swaggie—no trouble and strife tonight."

"You sound like a bloody cockney, Fraso. How are you and the wife getting on?"

Jesse struggled with a cowlick at the back of his head. "Better, thanks. I convinced her to stay home today. In exchange, I'm taking her to the theater this evening. How about you getting back to your digs at a reasonable hour tonight? Tomorrow's Ladies' Day, but we'll probably have to play our opening doubles match late in the afternoon."

Wick gave him a short punch on the shoulder. "Don't worry about me. You just go out and have a lovely marriage."

Jesse bent down for one last look at himself in the mirror, made a face, and withdrew with exaggerated dignity.

18

Laurie Silverman was number one in the United States and two in the world. Her pushy mother and gentle, handsome father expected her to be the first seed at Wimbledon this year. Consuelo Alvarro, the Margot Fonteyn of tennis, elegant and powerful, had injured an ankle in a home tournament in Buenos Aires, and her dubious fitness was a source of irritation to the bookmakers. She hadn't played for two weeks, missing the hard conditioning which was essential both mentally and physically for championship play. The bookies knew that Consuelo hadn't entered the singles at Queen's. They knew also that doubles are no test of singles ability or endurance, and the "books" were totally confused. The smart money said Laurie, if her mother left her alone (a gossamer hope at best, unless "Mom" was kidnapped).

Laurie inherited her elegance, good manners, and fine Levantine features from her well-bred father—nothing from her dumpy, rich, self-indulgent mother. Barry Silverman had been bankrupt twice by the age of thirty and married big-mouth, big-ass Sharon to bail himself out. He was a sensitive man and would have been far better off, as the saying goes, to "Borrow money from a bank—don't marry it. It's cheaper."

A perceptive father, he and Laurie shared secrets and love, and because of him she kept her sanity during the unnatural pressures from the twelve-and-under state tournaments on up the ladder of torment. Tennis mothers, and sometimes fathers, are a legend, a running joke among the cognoscenti of that strange world, but Sharon Silverman was a living legend; her scenes, intimidations, rantings, coquetry were famous (she once offered her body to a ranking official of a tournament to change Laurie's draw, but was politely thanked and her offer "reluctantly" declined).

Laurie walked down the hall of the hotel to her family's suite to say her good nights. It made her more nervous to do this than to play a match. The last two years had lifted the tension somewhat—playing on the Virginia Slims circuit, she saw her mother with ever decreasing frequency.

"Good night, Mom." She kissed her mother's cheek with a tinge of revulsion. Always that unnatural, sickening feeling when she was with her mother. She observed over her mother's shoulder what the British considered lavish, American, nouveau riche decor. "Venetian Jewish," Laurie thought, and chuckled to herself.

"So what are you laughing?" her mother asked, her eyebrows raised on her overly made-up face.

"This apartment, Mama. It looks like those pictures of private cars on those turn-of-the-century trains. You know what I mean? All these heavy red drapes. It's too funny. How can you stand it? I'd expect Vincent Price to jump out with a knife when you switch off the lights."

Her mother made a long face, as though her own personal taste were being questioned. "You chose this place because you wanted to be near your friends. I'll tell you something, Boobie . . . it's not my idea of the Waldorf either."

"Sure. I know, but they have lots of other types of accommodations in this hotel. Well. You're the ones who have to sleep in it. I just think it's very funny, that's all."

"Good night, Sparkle," Silverman called from the echo chamber of the shower. "See you in the morning," he shouted over the roar of the water.

"No, Dad. I'm going out to Queen's to practice in the morning. We'll go shopping in the afternoon."

"What?"

"Never mind! Mom, tell him for me, will you? This is ridiculous. It's like a giraffe having a conversation with Niagara Falls."

"Of course, sweetheart. I'll tell him. But am I included in your lives tomorrow?" she whined, her New York accent dripping with petulance. She had never lost her New Yorkese even though she moved to Fort Lauderdale shortly after her marriage.

"Oh, Mom! You told me you were going to visit Shirley in St. John's Wood. Did you just change your mind again?"

"Well, I'm sure your Aunt Shirley would like to see *all* of us." Mrs. Silverman flounced over to the small bar and began fixing a gin and tonic, heavy on the gin.

"Are you getting on that old 'I'm going to get loaded again because nobody loves me or understands me' routine? God!"

Sharon Silverman whirled on Laurie with a haughty, I-carried-you-in-my-womb posture. "Don't you speak to your *mother* like that! You are *not* twenty-one yet and I can still cut you off!"

"Mother," Laurie started the sentence in a low octave and increased volume and pitch in crescendo. "I made one hundred and eight thousand dollars last year, *do you hear me? You can't frighten me anymore! Take your money and fuck off!*"

They both stood, transfixed with shock at the last part of the outburst, which seemed to reverberate in the sudden silence of the room.

"Laurie! And when did you start that gutter talk?"

Flushed with a combination of embarrassment and relief, Laurie spun on her heel with a toss of her pony tail and stiff-legged from the suite.

She didn't go to her room but marched to the elevator. It was bad enough worrying about getting her period right in the middle of Wimbledon. She didn't need the nerve-jangling stresses of her mother's peevishness lurking just behind the curtains of her consciousness. The whole scene was so petty and unreal. Shirley was no more her Aunt than Clare Boothe Luce, and the day had already been decided, and "Oh shit!" she spat out loud in the empty lift. Thank God the Billie Jean Kings and Gladys Heldmans of the world had paved the way for her independence. She shuddered to think what horrible power and influence the LTAs and overbearing parents of the world had wielded in the past, especially over girls. She thought of the advertising phrase invented for Virginia Slims cigarettes: "You've come a long way, baby."

The doors of the elevator whirred open and she strode out of it to the nearby bar-lounge. It was pleasantly dim and subdued; a soft hum of conversation and a hypnotic piano chording unobtrusively in the background.

Laurie busied adultly up to the bar, not yet at ease in the role of being "of age." She was still confused by her scene upstairs and sudden metamorphosis. She stammered when asked for her order. "Ah— I'll have—I'd like . . ."

"She'll have a whiskey sour," Shep contributed from the next stool. "What's a nice little girl like you doing breaking training in a place like this?"

A flood of warm relaxation washed through her at the recognition of a familiar face, and she slumped onto her bar stool with a sigh.

"Old babycakes! When did you get to London, Shep?" she murmured, laying her hand gratefully on his. Shep glanced down at the cool touch of her hand and tried to give the action an instant evaluation.

"I've only been here long enough to unpack and practice at Queen's. Would I be nosy if I asked you what put you into your apparently awkward state?" Shep asked, taking his drink with his left hand so as not to disturb the warmth the other one was enjoying. He had taken his degree in clinical psychology and was thus torn now between the roles of father-confessor and seducer. They often complemented one another, actually. He reversed the position of their hands, taking hers firmly in his right hand. With a deft motion, Laurie removed her hand and patted his with a "thanks-but-no-thanks" tenderness.

"I'm all right now, thank you, Shep. You wouldn't take advantage of an agitated little girl, would you?"

"Well, Miss Silverman," he answered, spinning toward her, "yes. Yes indeed I would. You are very nearly twenty-one if you aren't already and, oh, I guess about five foot ten."

"Shep, darling, you know from all the magazine profiles on me that I'm *exactly* five feet ten. I think you're divine, but having an affair with you would be like repairing a watch in a mix-master."

Shep laughed a genuine laugh. He sipped his drink, before continuing. "I'd be very gentle with you"—this in an exaggerately "older man" theatrical reading.

"Dear Shep . . ." Laurie's dark eyes and bright, good teeth caught the indirect barlight with sharp sparkles. "You are a dirty, middle-aged man and just because I'm overwrought"—she was completely relaxed now and enjoying the light banter—"I'm not going to permit you to have your wicked will with my pink, tender young bod'." Her allusion to his being middle-aged stemmed from the fact that his rich, chestnut hair was indeed slightly receding. In fact, the remark was tinged with just enough truth to hurt him minutely and fleeingly. She noticed it, but didn't want to aggravate it by trying to qualify.

"Laur, you make me sound like some sort of aging, lecherous surfer," he remarked softly.

"Oh, for heaven's sake, I didn't mean it. By the by, have you ever noticed that all surfers seem to look like a tall Alan Ladd with long hair?"

Shep laughed with a swift mood change. They did look like Ladd. Laurie fiddled around with her drink and went on, changing the subject. "I just had a doozy of a row with my mother over nothing . . . but I broke fresh ground in our relationship. I've been faking it for years, and tonight I told it like it is, to borrow a dumb phrase. Not with much finesse or brilliance, I admit. I told her to eff off." She said the last vacantly, into space, trailing it off to a whisper.

"Well, little bird, I suppose it was the place and time or whatever. Why don't you finish your drink, skip rope one hundred times, and get to sleep. Don't dwell on it. This is no time to let peripheral things crowd your concentration. You've never won this big number and you've got a great chance this year."

"You mean because Consuelo's hurt?"

"No. I mean on your current form you could whip her anyway."

"Missy Teaford's looking very sharp, too. Oh, Lord, Shep. So many of them are reaching their peak performances now. They are really on. It's just a question of the breaks."

Shep chuckled over a sudden remembrance. "I was just thinking of the time Missy was playing a junior event on an outside court and forgot to put her pants on. She was so nervous dressing, she simply forgot and ran out to her court." Laurie smiled as Shep went on. "Laver and Okker were playing on the first court, and I swear there were just as many people watching Missy." Laurie called for her bill. "She even won the first set 6-2 before sombody told her," Shep laughed. "Then she ran off the court crying."

"On that sad note I'm going to bed," Laurie said.

Shep reached for her bill.

"No, no! I'm a liberated young lady now. I made almost as much as you did last year." She signed the bill and danced off to the elevator with a good-bye flutter of fingers, feeling very, very glad she'd seen Shep just at this moment.

Throughout the 1970s women's tennis had become increasingly popular, perhaps because it offered the type of play with which most people, men and women, could identify. In men's tennis, the sheer speed with which each point is decided takes some of the drama and suspense out of the game. In women's tennis, long rallies and more chess-type strategies made it great fun to watch. While most club players, park players, and beginners aspired to the split-second reaction time of the men's stuff, they had neither the talent nor conditioning for it. For the most part, women's tennis was their tennis, although, except for the very best club players, the women would knock their ears off too.

In short, the women were just as capable of pulling in the crowds as were the men, and at Queen's Club on opening Tuesday, traditionally Ladies' Day, the crowds were there.

Angie Redfield was miffed at not having been seeded. Not that she deserved to be. Her record for the year was dismal. She had dropped off the circuit twice to oversee the organization of an indoor complex in Minneapolis, where she was to be resident pro, and her sex life was taking up too much of her back-up energies. Since her public *coming out* in Austria, she was now duchess of the sapphic set. No AC/DC for her anymore. She had made her choice, and was very relieved to be no longer wallowing in the miasmic sloughs of confusion. For whatever the reason—psychological, biological, or both—she preferred women.

Now, first round, here she was, up against the first seed—Laurie Silverman.—Goddamnit! "If there weren't any bad luck, I'd have no luck at all," Angie thought, not even trying to be philosophical about it—after all, here was the chance to play the best player of the moment, an excellent opportunity to test herself. They weren't even on court one, where you could reasonably expect the first seed to be, but on court two. Missy Teaford, a former Queen's winner, was playing the English number one on the main court.

"Great! That's just too terrific," Angie spat out as they walked out to court two. "Don't you—I mean—aren't you a little pissed off to be shoved out here?"

"I don't care where I play on the first day as long as I'm on court one on the final day," Laurie said cheerfully. They didn't like each other, but tremendously.

The match went as one would have expected, with Laurie's big serve and Evert-tight ground strokes. The first set went to her 6-1. In the second set, Angie put up a real fight—looked like a terrier zipping around the court—or like a shortstop moving precisely to left and right with no wasted motion, and she had chances to break Laurie's service. Relaxed by the easy first set, Laurie's first serve wasn't inches off but, at times, feet. Few things are more unsettling than a serve gone awry. It upsets the rhythm of the rest of the game. She was just managing to hold serve by spinning the second one in and staying back—hardly a grass court game. It put heavy pressure on her strokes, and she wasn't enjoying it at all.

Angie kept hanging in there, contesting calls, talking aloud, occasionally being rude, and looking quite butch. She thought, "If I can win a set, I can win a match."

At 4-all, 30-all, she returned Laurie's serve deep to the corner. The ball carried out about three inches. Laurie made no attempt to hit it —but there was no call. She stopped rigid and stared at the linesman, who called "Out!" belatedly. Angie rushed up to the net, her face a mask of rage. "Now goddamnit! You stop intimidating those linesmen—you're cheating again, you big Jew! Once more and I'll light the fuse on your Tampax and blow you to smithereens."

"Don't talk to me like that, you little bull dyke. I'll slap your queer face!"

"Ladies, ladies," the umpire said, the words oily with sarcasm. "Play will be continuous. Now, we don't want anyone disqualified, do we?"

Laurie was seething and in her transformed state of mind, her serves bombed in time and again. She went for all her shots, and Angie was blown off the court. Laurie ran around the second serve, hitting straight-out winners, and the match was in the books ten minutes later, 6-1, 7-5. As they left the court, signing autographs as they went, Angie said, "I'm sorry about what was said out there in the heat of the moment."

"No reason to be sorry. Nothing was said that isn't true. I *am* having my period, I *am* a big Jew, and you *are* a little bull dyke. But one more word and I'll slap you silly. If you want to be a man, go to Charles Atlas." With long strides, she left Angie far behind.

On court number one, Missy looked cool and elegant, her strokes graceful but crisp. She knew the mathematical, the geometric pattern of the court and used it. The nervous British girl, anxious to do well in front of the home crowd, was trying too hard, her two-fisted backhand and forehand flaying rather than stroking. She never played well in England. Her parents had forced their little Belinda, a natural left-hander, to learn right-handed; then, with later enlightenment, permitted her to switch back. Now, in a tight spot, there were patches of indecision, barely discernible, but enough to bungle a shot. She made a good showing today against the second seed, going down 6-4, 6-4.

In the tearoom, Missy saw Shep in a deep discussion with Phil Katz, his financial advisor. She felt nothing. No bells rang—no twinges—no pleasant memories—no torches—the total nadir of emotion. Objectively, she noticed he was attractive. On the other hand, she also noticed that Phil was fat and handsome, and that it might rain.

Queen's was her first major tournament since resuming tennis. She had played two small ones in the States, but mainly she had gone back to the fundamentals. Day after day she hit hundreds of balls—all strokes—smoothed the hitch out of her serve, and developed a reliable sliced backhand down the line, a stroke she had never really possessed before. She read that Lew Hoad had won match upon match from Pancho Gonzales because Pancho had had the opposite trouble: he *always* hit his backhand down the line with boring consistency, and Lew would follow it in, jump across, and powder the volley cross-court far out of the reach of Pancho. Then Gonzales went to work by himself, and in the middle of his career altered his shots, in fact learned a brand new cross-court and turned the outcome of the matches around.

Missy thought, "if Pancho was willing to do that at his stage of tennis, there is no godly reason why I shouldn't." A story she heard about Gonzales confirmed her belief in the new positivism, the Transcendental Meditation she now found strength of mind in. (Her mind was growing healthier than it had ever been before.) As a boy growing up in a poor sections of Los Angeles—she recalled the story—Pancho was the second-best marble shooter on the block. He won marbles by the dozens from everybody else, only to be cleaned out by a black kid—who had no arms; he shot with his toes. The story goes that Pancho said, "If that cat can whip me using his feet, I can be the greatest tennis player in the world." And there can be no doubt that he was.

That was the type of concept Missy reached for when she was bored or frustrated learning a new shot, or simply dragged practicing or skipping rope. It added a half hour to her work and a hundred skips—willingly. One of the reasons she maintained a friendly relationship with Shep was not to waste energies in reliving former traumas. Strangely enough, she avoided Wick for the same reason, but it was Wick's energy she wanted to conserve; not have him dissipate it on foolish defense mechanisms. Wick wouldn't believe her, but his secret was quite safe. No one need ever know that, with all his bluster, he was completely impotent with women. At first, Missy was afraid that it might have been her fault—too much fondness, too much desire for him. An Australian girl, a non-player, told her in confidence once in Melbourne that Wick had been unable to perform with her. He said that it was a mental block he was adjusting to.

Missy was aware that he said ugly things about her, but knowing why softened the thrust. She didn't want to cause him any more anguish than already plagued him. She, above all, knew the pain of mental confusion all too well.

Not only the abortion of Shep's child, but, long before, the combination of the impercipience of her parents and the onion-skin layers of pressure—first classical piano, then the horrors of junior tennis—had caused Missy to recoil from the world. She had found a small, safe, quiet place in her mind and went there to live for a time.

In retrospect, Shep had been fairly decent about it—the money—his presence, but he was very young then, hardly the man of solid priorities and style he was today. Compassion and sensitivity were gaping holes in his personality makeup. Missy, at the time nineteen years old, went through the agonies—lost, frightened, confused, suicidal, and, more than anything else, alone. He wasn't there when she needed him; not in spirit. Nor were her Victorian work-ethic, red-white-and-blue parents. Missy hadn't, in fact, yet come to terms with the question of abortion, the profound moral considerations of the act. Is it murder? Is the embryo a person, etc.? Now it was academic—but then—then it was the most paramount, insoluble problem in the universe. And the odd thing about it was that her parents, her *American Gothic* parents, urged, insisted, that she abort. How about them apples, Billy Graham?

Missy took her tea and sandwich and decided to sit with Shep after all. They, at least, never had to make an effort.

"Hi."

"Well played, Miss. You looked fine. Mighty good out there."

"Oh, thank you. What was all the ruckus on court two?"

Shep took a bite of his scone. "The little birdies were fouling their nest. Evidently, Laurie Silverman didn't take any lip from Angie, if you'll excuse the expression. These scones are hot and delicious with butter. By the way," Shep assumed his professorial attitude inherited from his father, "did you know that you—that is, 'one'—doesn't have tea and crumpets anymore in England? Crumpet means pussy now. Not the thing itself, though. That's called fanny. That's right, a fanny. Crumpet is more abstract."

"You're putting me on, Shep."

"I'm not—so don't go around talking about anybody's big fanny. Especially mine, 'cause it isn't nice." Shep glanced at her sideways, appraisingly. "You look very well, you know—pretty—adjusted."

"That's very nice. Thank you. I'm in good shape."

"Nutty life, isn't it. I'm glad we're still friends, after all, Missy."

"Are we? Or are we just too lazy to be enemies? I wonder. Pass the salt and pepper, please."

"Listen—I've been sorry about that for so long, I'm numb from being sorry. That was almost four years ago and, although I can only be sorry in the helpless—what can I say? whatever it is—the way a man can be for such a thing. Oh, I know it was ten thousand times worse for you, but I *can* forget. Won't you?"

"It was one of the major things that adds to the total of what a person is. To try to forget is a denial of yourself, and that's surely a form of bullshit."

"Well, that's all very complicated, I'm sure. I thought for a moment there you weren't going to find your way out of that sentence. I have to take off, sweetie. My doubles is next on. Want to have a drink later?"

"Shep, I swear that's what you asked me four years ago. I wish I'd said 'no thanks' then. No thanks. I'm busy, and if I'm not busy then I'll get busy."

"Okay. I hope Mr. Right comes along for you."

"Shit!" she said demurely. Missy surveyed herself mentally. Better and better. She guessed that she would be all right. As she left, she said some encouraging things to Belinda Martin. If the mass of men and women lead lives of quiet desperation, then every little bit of encouragement is a plus. Missy had decided that bittersweet was a big improvement over no sweet at all.

Angie stomped in with a posse of multishaped, international lesbians; some tennis players, some show business celebrities. It was bizarre to watch the petite Angie trying mannish gestures. She did everything but wear a derby and smoke a cigar. Inge, the Swede, *was* smoking one. Perhaps it was only a matter of time. Missy went out to the gallery to watch a South African girl who had won

Junior Wimbledon the year before and was causing quite a stir in the tennis world. It was always stimulating but also frightening to see the new girls come along, waiting in the wings to pounce, to have their time.

Pony tails or curls or Indian headbands, wielding racquets with God-given talents, their faces frozen into squinting seriousness, melting only rarely for the odd smile as if humor might destroy concentration or be misconstrued as immaturity. All sizes and shapes and colorings, yet identical in their youthfulness—and behind them, just beyond, another crop ready to cannibalize them in their turn.

Anyway, goddamnit, it was still Missy's "turn" and she was going to avail herself of it. "It's *my* year, you young duckies, and don't you forget it. Let's have a little respect around here!" The thought made her smile. It was almost a supplication.

The South African had a double-handed forehand but an ortho-dox backhand. She was a small, lovely blond thing whose overall impression was marred only by the constant grimace of determination. The lines invariably became firmly etched by sun and weather in later years, giving even the most pleasant personality an angry visage. But when she smiled, this little Maryke Tamboezer, it was like the sun coming out—eyes crinkled, teeth shone. Missy knew she was watching a great player of the future.

Billy Sherman walked by. "All set for your doubles, pretty Miss?"

"Like a tigress."

"Isn't it funny that when dogs show their teeth we know fear and when peole do it we feel good?"

"Yes, Billy, very funny indeed." Thus he shared his philosophical diadem of the day, and ran up the stairs three at a time.

Karen Fraser came out of the tearoom and, having no other way to go, had to say hello to Missy. She had even considered faking a phone call on one of the three phones on the wall. Why they shared this antipathy, neither ever knew. They had been introduced only once, had never heard anything bad of the other, nor had they had words. But like strange dogs, their nape hair frizzled on their necks when they were near each other. Missy thought it a good time to lay the ghost. They were going to be around each other for almost three weeks, and if at all possible, Missy thought, one more point of irritation, so unnecessary in life, might be neutralized.

"Hello, Karen, Good to see you. You're looking well. You've lost weight, haven't you?" (Bad start! The inference could easily be drawn that she was fat before.) "I mean—we all have a tendency to get sloppy." (Worse.) "That is—I do. To tell you the truth, what I want to ask, actually . . ." Karen was slightly amused, waiting for the other shoe to drop.

"Why don't we seem to get along?" Missy said. "Was it something I said? Or did? I hardly know Jesse, so that can't be it."

"God, I don't really know, Missy. How very funny of you to mention it. I never thought you liked me, I guess. You know—bad vibes and all that?"

They sat down together on the club veranda, crossing their legs in the same direction like the June Taylor Dancers. They noticed it and that brought a short laugh—a wedge into the wall.

"While I was 'resting' at Silver Mountain in Massachusetts, one of the doctors was interested in parasensory perception," Missy explained. "You know what I mean, telepathy and all that jazz? Well, they knew that thought waves weren't transmitted by electricity, because the amount of power generated by the body to work it and to think, is tiny. It can only be picked up right next to the skull. Infrared, he figured, might be the answer."

"What does that have to do with us?" Karen asked.

"I'm getting to it. Everyone has an aura. They proved it. Really. Some people's aura went as far as fifty feet—most, though—oh, I guess it averaged out to about five. The way they measured it was to have a person stand near or against a wall for five minutes—then step aside and they would take a picture of the wall with an infrared camera. The heat of the body changed the molecules, the molecular structure of the wall at that spot, or hotted them up or something. And there was the outline of the person on the film. Then they had us—I was getting mine done, too—something to do—they had us move forward, still taking the pictures, until we didn't reigster any longer. It wasn't any good for telepathy, even fifty feet is useless, but what it did teach us is that two people's auras sometimes give off negative charges when the fields of force come into contact. I mean, there are two perfectly innocent guys ten feet apart not liking each other because of their stupid auras."

"And you think that might have been what happened to us?"

"You said might *have* been, right?"

"You're quite right. I see what you mean. Let's have a drink later, after your match, and talk about it. I don't know enough about those things—telepathy and everything—and they fascinate me."

"Sure! But tomorrow. We'll leave our auras at home. Ciao, Karen." One job done well. Missy went off to her next project with a chest full of confidence and goodwill. In a lopsided contest, she and her partner, Anne-Marie Dubois, gave an all-French pair a great waxing. A well-rounded, healthy, full day in the recovery of Jane "Missy" Teaford. With her copy of *Psychic Research Behind the Iron Curtain,* she left the tennis beehive for some solitude—a positive solitude, a peace of mind that was reassembled in richer condition, salvaged as it was from a shattered mental mirror of horrors.

J. C. Sark of the *Evening Standard* was taping a brief interview with Wick in the Queen's Club billiards room.

"Alvin Wick—do you mind if I call you Alvin?"

"Outside of my mother, you'll be in a class by yourself. Al or Wick, if you like."

"Well, Al, you'll be going on court in a few minutes against the great Franco Berconi. How do you appraise him compared to past great players? And . . . eh . . . how do you fancy your chances?" Sark wore rumpled, baggy tweeds under his rumpled face that hadn't been shaved recently. It was his trademark. He wrote rather well, though—not in the class of, say, David Gray or Tingay, but lucidly and with fairness.

Wick hadn't wanted to grant the interview just before going on for such a big match, the quarter-final, but he had put Sark off as long as he politely could. He now, however, was warming to it, in only the way that a microphone can persuade one. Even the timbre of his voice changed and purred with a didactic, for publication, tone.

"Not only because I'm about to play him . . . my chances aren't any fifty per cent, in answer to your second question . . . but because, and the boys will back me on this one, because Berconi ranks up there with the *big* ones. He's every bit as good as, say, Nastase was in his day . . . you know, the natural flair or whatever—in fact he reminds me quite a bit of Nasty . . . in temperament and style. I wouldn't be at all surprised if you asked him, if he didn't tell you that he patterned himself on Ilie. He was a kid when Ilie was at his peak and, I'm sure, identified, like, with him. Berconi is even nuttier, though."

"Do you think Franco can maintain this feverish pitch of play? I mean . . . he is playing the most superb stuff, isn't he?"

"True. Even Laver at his best would have had trouble with him. But I'd like to say, Laver was the best and would have taken him eventually—too much raw power mixed in with all the shots for Berconi. I'll tell you something. At this rate, Franco's gonna be snapping and biting himself next week at Wimbledon."

"Who do you like at Wimbledon?"

"Outside of myself, of course, Berconi, as we've said, looks a good bet. Jesse Fraser on his day is in with a chance. Francis Shepard looks very sharp. Even your own British boy, Jaguar Gray."

"Is Berconi liked by the other players? How do they get along with him? You know . . . after all . . . he is a prima donna and a bit off-putting on court."

"Well, there again you have to compare him with Nastase. He'd drive you up the wall during a match, but afterwards . . . you know, the guys actually like him. Strange thing is, for all the pressure, and sometimes bad feelings that pop off on the court, the people on the circuit get on amazingly well. We're pros and when the match is over . . . calls and net cords or psyching or whatever . . . when it's over it's over and one has only oneself to blame for losing. No. We get along fine. Except for one prick—Oh! Sorry! Can we take that out?"

"Don't worry, I'll fix it. Who is that, then?"

"I won't mention any names but his initials are Randy Mariano.

He's bad news! Berconi's won Wimbledon before and so has Jesse Fraser, so they'll be more relaxed. These guys are as good as, or in some ways better than, the people before them. Like track records keep getting broken. The first time I played Wimbledon, I came up against Jimmy Connors in the first round. Both of us left-handers. I never knew what hit me. But I'd like another chance at him now. If he were still playing. Hey, that's me. I hear them calling me on court. So long, Sarko."

"And thank you, Al Wick, for an inside view on the coming Wimbledon. Good luck."

Wick won the toss and elected to receive. That started Berconi out relatively cold and looking up on the service throw into the two o'clock sun. It wouldn't bother Wick. Serving left-handed, he looked the other way. A slight breeze occasionally rippled the Queen's Club pennant; jackets were coming off in the sun-swept stands, and a pleasurable afternoon bubbled in the sweet calm that preceded the first serve.

"Players ready? Linesmen ready? Play!" the rich British accent of the umpire intoned. Wick was at the peak of his playing career and had been playing up to his best all week. He kept his racquet unusually low to receive service because they skidded low off the grass and shot into him. No use wasting time lowering the racquet. With two loud cracks, he whacked clean winners off Berconi's first two serves. It was his style to go all out, and when he was on he was tough. Berconi smiled and clowned around, importuning the crowd to recognize that Wick's shots were nice, but sheer luck. Franco hit a backhand volley cross-court but short, and Wick, running wide, thumped a forehand down the line. It went by the incoming Berconi like a taxi on a wet night. At love-40, Berconi sliced a serve down the center that Wick barely got a racquet on. The backhand went off the wood and dribbled over the net. He had a service break. All big Australian grins, Wick was wiping the grip on his racquet behind the umpire's stand. One of Berconi's tricks was to talk to the opposition on the changeover. Generally, it was considered rude and unprofessional to break the other man's concentration and to influence his play by talking, and most players were silent and tried to ignore Franco when he went into his act. Not Wick. He was born to banter. It was his nature.

"Goddamn . . . wake up, Weeko . . . you gotta *bella fortuna*. Lucky, you justa lucky. You oughta be ashamed with that net cord."

"Yeah, Franco. It's like eating pussy. I like it, but I'm not proud of it. Anyway, things even out, right, mate?"

Wick won his serve with the loss of one point to lead 2-0, but from that moment on it was all Berconi in the first set.

From the window of the dressing room, Jesse and several others were watching the excitement.

"Chroist!" Jesse said, with a deep sigh. "I've never seen the Wicko play so well, and still Berconi's murdering him. No doubt about it, he's something else."

Shep nodded agreement. "Wicko's not giving much away like he usually does, but I'm not sure that's the way to play Berconi. He lives off the other guy's pace. Wick doesn't mix it up enough."

Shep turned from the window with a shrug. He was thinking ahead to Wimbledon, and once more the tightness gripped his intestines. He wanted to win so badly that it was like a disease—a very painful disease. Ambition and achievement—the driving forces in life, bearing along with them anticipation and lurking disappointment. It haunted them all.

Jesse too wanted to win at Wimbledon this year with his whole being. He had secretly decided to retire at the end of the year to make a less erratic home life for Karen and their little girl, Zoe. The child had been named after Roger and Frances Taylor's daughter; they thought that if their baby turned out as well as Zoe, she would be a real charmer. Jesse knew that if he wanted to save his marriage, rocky as it was, he would have to be there himself to do it. Long-distance calls didn't suffice anymore.

"What's the score?" he asked. Freddie Moore was hanging out the window. He turned and yelled, "Berconi ran out the first set with six straight games, but Wick has a break in the second." Some fans shushed him from down below. Freddie pulled back from the window, embarrassed.

On the court, Berconi, confident now that he could break Wick whenever he wanted to, was showing off, toying with Wick, blunting his power with touch-stop volleys off sizzling cross-courts—lob volleys just over the head and out of the reach of Wick. It was brilliant tennis, and the crowd loved it. A lesser opponent might have been broken by Berconi's arrogant virtuosity. But Wick just kept blasting away. His margin of error was low, and today he was homed onto target. Somehow unperturbed, his game flowed. With nothing to lose, he "let it all hang out." Relaxed and moving well, he held onto his serve. Several times there were break points against him, and Eduarde Vector's voice would sound clearly down the years: "the ball, play the ball, not the reputation." Time and again he'd crunch a backhand volley to save a break or win his serve. It had become a point of pride with Berconi to pass Wick on his backhand side, and the shot wenteth before a fall. Berconi didn't break, and Al Wick served out the set.

The crowd warmed to the possibility of an upset and people crushed into the stands for the third set. Some of the steam whistled out of Wick in the third set. It was as though winning a set had been accomplishment enough against the world number one. Franco Berconi went quickly ahead for a 4-1 lead, once again the master. On the changeover he was once again his sparkling, arrogant, annoying self.

"You played good, keed. But now you taka you medicine. She's a wrote it all."

"Blow it out your ass, Caruso." Wick wasn't angry at Berconi, but disappointed at his own letdown; his own negative reaction. He was frustrated.

"I won't change anything, goddamnit! I'm playing bloody great and he has to beat me. Attitude," he thought, "attitude."

Attitude or no, Wick was down 15-40 on his serve—Berconi had two points for 5-1.

"What a shot!" Shep screamed. "Wick just hit a backhand drive-volley on the baseline. Berconi's shot had him beat off the ground! Oh, oh. Now it starts."

Berconi had his hands on his hips, glaring at the baseline judge, who sat hunched over, his outstretched hands palms down, his eyes riveted to the spot of contact.

"*Hai bisogno di occhiali, stronzo?* You were asleeping, *davvero?* Shit!" Berconi started strutting around like a gladiator, appealing to the crowd for justice. They hissed and whistled, some of the ruder ones shouting:

"Get on with it, Macaroni!"

"Come on, you big spaghetti bender, play tennis!"

Berconi put his fist in the crook of his elbow and clenched his other hand. "Fucka you! How you lika punch in de mouth?"

"Play will be continuous! Come on, Berconi!" the umpire said with exaggerated patience. "The ball was good, 30-40."

Berconi crouched low to receive service. Wick felt the electricity, the sudden change of polarity. The tempo had definitely dropped a beat. He served down the middle and stayed back. A surprised Berconi pushed a weak, mis-hit forehand into the net for deuce. His dark, curly hair was matted flat on his head now. A small titter of laughter followed in the wake of the mis-hit, and he was plainly disconcerted by it, looking darkly into the stands. Wick held serve.

At 4-2, 30-all, Wick put up a low, deep lob over Berconi's backhand. Leaping and twisting in a spectacular bit of ballet, Berconi smashed it deep into the left corner, rifling chalk a foot into the air.

"Out!" came the call.

"30-40," the umpire droned funereally.

Berconi turned toward the clubhouse and hit the remaining ball in his hand straight through a window, slightly cutting the forehead of a girl player inside with a shard of glass.

"Restrain yourself, Berconi. Are you sure, linesman?" the umpire asked.

The linesman, the same one who had made the previously contested call, stood up from his chair, walked over to the baseline, and stamped his foot on a spot he was pointing to, just behind the line. Chalk again flew up, fortunately for him after such a dramatic gamble. People were calling "Out! Out! Out!" from the stands, with a small sprinkling of "Ins!" The outs had it.

"Ladroni!" Berconi shouted, "you are stealing from me, you *bastardi!"* He walked over to the umpire's chair, actually foaming. "I want thisa guy off!" he hollered, wild-eyed, pointing at the linesman. Wick had sauntered into a shady place in the back court and was sitting quietly. A new look of fatalistic confidence was apparent on his relaxed but ready face. He was prepared to play his part in what had become an athletic farce.

The umpire sent a ball girl to get the referee. The referee, a pretty, middle-aged woman named Lola Holguin, was already on her way down from the terrace and walked on court with a smile of genuine good humor. Berconi immediately figured she was in the plot.

"What's the problem, Franco?" she beamed.

"Ma, Lola, lei parla l'Italiano, vero? Quel cretino ha fatto due errori critici, e non voglio che resti in campo."

"Ti prego, Franco, parla in inglese—the umpire doesn't understand."

"Di lui, non me ne frega niente!"

"Well, I do, now—the linesman stays. I saw both calls and agree with him. He's a veteran umpire, an honest man and I think he should stay. On the other hand, if you don't behave, watch your language, and clean up your whole act, I'll disqualify you. Now, I'm sorry, but there you are; that's how it is."

"You not-a my mama. I don't have to take this kinda talk from a women!"

Lola's smile had long since disappeared. "At this moment, I am the tournament referee—and as such, I won't stand for any of your nonsense. Now get out there and play!" She spun around and, head back, strode off.

"30-40," the umpire said crisply into the microphone. Berconi walked back to the service line and angrily served two first serves in a row into the net, double-faulting Wick into game.

Like a vicious terrier, Wick got a hold on the match and, with new energies, kept a grip on it. The psychological momentum was, of course, all his. Berconi was playing emotional tennis—sometimes brilliant but not percentage. He started playing Wick's slugging style, and put on a great show of enjoying himself. He was all smiles and compliments. Wick turned the conversation off in his mind. He had the match in his hand and he damn well knew it. No! He wouldn't be talked out of it. Two bad volleys, a really bad call, and an unlucky shot by Berconi, and Wick had a break for 5-4. Berconi didn't query the call—just fluttered his fingers at the sideline judge, whose ears were crimson and whose eyes gave the unmistakable impression of belonging to a man who wanted desperately to be someplace else.

They both walked behind the chair. Berconi poured a paper cup of orange squash and sat down. Wick toweled off his racquet and face, and glanced briefly up at the faces lining the dressing room windows. No one was dressing. Every window was filled with players watching Wick on the verge of his greatest win. He didn't linger, didn't focus on anyone. With purpose in his step, he walked firmly out to serve. If Berconi said anything, Wick, in a trance, did not hear it.

"Can he do it, Jess?" Shep asked.

"Wouldn't it be great? Yeah! Why not? If he doesn't get the shoulder, he's got him. I've never known Wick to choke."

The first serve whistled down the center to the backhand and Berconi chipped it into the net. Wick tried again to win the point outright and swung the serve wide to the backhand. Berconi took it early, finessing it cross-court for a storybook winner that Rosewall would have been proud of. Upset by the artistry of the shot, Wick hurried his delivery and double-faulted: 15-30. Wick bounced the ball three times, reared back, and uncoiled a blazing ace past—the forehand. Berconi tucked his racquet under his arm and joined the applause, clapping and bowing slightly. The next point went: serve—backhand up the middle—volley—lob—smash—lob—retrieve—and finally a smash that bounced into the stands. Wick was at match point. "I'll just get a medium-paced first serve and hope he goes cross-court. Because that's where I'm going to be. He's been trying there all day," Wick thought, in the instant he bounced the ball—three times. The only sounds in the immense compound came from several clucky pigeons on the roof of the clubhouse, and the bouncing of the ball.

Berconi wasn't fooled by the change of pace of the serve, but got a slightly bad bounce off the grass. He was already committed to a cross-court return and perforce had to finish the shot. The poised Wick, in his hot-eyed eagerness, almost put the volley out, but got a piece of the line for—

"Game, set, and match to Wick, 2-6, 6-4, 6-4!"

His racquet sailed up as high as the clubhouse roof, sending the flock of pigeons into sudden flight—this time they were unheard over the tumult. Wick rushed forward, his hand out to meet the smiling Berconi. They shook hands and Franco removed his shirt, in the tradition of football captains after a game—Wick started to pull his off.

"By Christ, he did, it," said Jesse. "The wild bugger won himself a great match. I'm sorry for Berconi in a way, but Wick really had himself a day."

"Hey, you guys!" one of the people still by the window shouted.

"For God's sake, come back here and look at this!"

Berconi had taken off his shirt and shorts, kicked off his shoes, was tugging at his socks, all the while smiling at the transfixed Wick, who was rooted to the ground, his own shirt in his hand.

Next, Berconi stepped out of his jock, while the umpire, halfway down from the chair was saying, "Now see here!" Berconi tossed his jock high in the air like a victory salute and started off the court, passing two schoolgirls in straw hats who were still trying to get his autograph. He made some headway through the stunned crowd before three bobbies from the gate bulldogged him into a Rothman's banner. He punched one of the cops right from underneath his helmet, but with the help of a few officials they managed to wrestle him into a storage room near the indoor tennis courts.

Eduarde Vector made his way through the agitated, laughing crowd to Wick's side. Wick seemed half anesthetized as he gathered up his racquets and gear.

"Hey, Mr. Vector. Nice of him to go 'round the bend and bugger up my win, wasn't it?"

Vector threw a towel around Wick's shoulders, aware of the cameras that were clicking all over the place. Several newsmen kept attempting to shove mikes in front of him as Vector led him through the mob.

"Not now, gentlemen, not now!" Vector said.

"Al, did at any time, did you . . . ?"

"No please—not now . . . later, press conference—later—let us through, please—excuse me—thank you—let us through."

They made it through the melée, which had more of the flavor of a boxing bout than tennis, and into the clubhouse. The players gathered in the downstairs reception room gave him loud applause. That brought Wick around again and he waved warmly in acknowledgment. Upstairs it was the same. Jesse threw his arms around his neck and Jaguar pumped his hand. "Fantastic!" Jesse said.

"Nice time for the bloke to go balmy, wasn't it?" Wick smiled wanly. "Why couldn't he have done it in the first set and saved ten years of my life?"

Poor Franco, Shep thought; he knew that much of the jubilation for Wicko was the manifest reaction of the anger and, yes, frustration pent up against Berconi over the years. In truth, Shep was relieved too. Berconi's outrageous display effectively put him out of

Wimbledon and raised Shep's own chances sixty per cent. He was annoyed and disheartened at the selfish thought—ashamed of thinking of his own advantages at some other poor shit's expense.

"Great effort, Wicko, really. Never mind about what happened after—you won the match. That's the thing." Shep shook his hand as he started out for his own match. "How do I follow *that* act, eh, Wicko?"

"I played all right, but I had all the breaks too. That last serve never came up. Dead bounce," Wick minimized. But inwardly he knew that he had just played one of the best matches of his life. And Berconi's post-match antics in no way diminished the brilliance of that win.

Jesse said to him as he walked to the door, "Remember all your good shots! A couple of shots don't make or break a game!" Both of them knew in their hearts that, all things begin equal, the breaks certainly did affect the outcome of a match.

"Jess, I'm going on back to the flat," Wick called after Jesse. "Bring the wife and let's have a drink tonight at the Gloucester with some of the mates."

"Right you are, Wicko."

Wick indulged in a long, pampering shower, luxuriating in the jets of water on his face, dressed slowly, and went to the press conference. All the predictable questions were asked and answered— predictably. Given the strange circumstances, the conference was unaccountably brief. Put it down to the finer sensitivities of the British press and the tragedy of Berconi.

On the way down the steps, a reporter who hadn't been at the conference ran up the stairs into Wick.

"Oh, pardon. Congratulations, Al. I guess you've heard that enough already today."

"Never hear it enough. I don't win the big ones that often."

The reporter kept looking over Wick's shoulder while he was talking. "I am covering Shepard for the *Globe*. He beat Owibakum, the Nigerian number one, 6-0, 6-1, in forty minutes. Have you seen him?"

"Owibakum?" Wick asked, wide-eyed.

"No, no! Shepard," the reporter said with grinning exasperation.

"The way things are going today, he's probably upstairs jerking off."

Wick signed a few autographs and drove off for some quiet reflection.

The room was full of tennis players but they were subdued and reserved in keeping with the quiet decorum of the Gloucester lounge. Wick had difficulty at first locating Jesse and Karen sitting in the pleasant obscurity of the piano bar. Karen waved, and Wick made his way through the tables with their orange-shaded imitation gas lamps, snapping his fingers to an upbeat tune the piano was knocking out.

"Well, you big pain in the ass. How's it going, champ?" Jesse greeted him. Wick leaned over and kissed Karen gently on the cheek. She was, like Scarlett O'Hara, "not beautiful, but men seldom realized it." Her movements were precise and graceful. Irish-Indian-French—flashing blue eyes over high cheekbones when she was angry—her large, braless breasts firm inside her Mexican blouse. But it was her coloring and skin texture that struck one at first. Raven hair framing the blueness of her eyes and reddish brown skin still gave Jesse a shudder of intense pleasure every time he saw her. She turned him on like a Christmas tree. He always feared losing her, but she could be a harridan too. The glamour of traveling on the circuit had long since jaded, and she couldn't watch Jesse's matches without becoming a jangling jello of nerves. For Jesse, she wanted this Wimbledon. But afterwards, for God's sake, some normality in their life, for all three of them. Perhaps four.

The conversation was sporadic, revolving around Wick's win and the phantasmagoria of Berconi. Jesse topped up his glass with the remainder of the Tuborg. Wick signaled to the waitresses. "Another round here and a lager for me, too." Karen often drank that little bit too much that the Indian and Irish part of her couldn't handle. Jesse had a theory that the Irish, Russians, Poles, American Indians, and anyone else on a mainly carbohydrate diet had a low tolerance to alcohol. Thousands of years with very little protein, just potatoes and corn. They even made their booze—poteen and vodka—from the potato.

Jesse said to Karen, "Did you know that the average life span in ancient Rome was twenty-two years?"

"Terrific," she said sarcastically, "I've made it five better, so I deserve another VO, right, Wicko?"

"Quite right, Karen. You get sloshed for all of us, we have three semifinals tomorrow between us. Just like old times, eh, Jess? What happened after I left the club?"

"Sherman and I did our usual thing. I routined him. I guess you know Shep had a breeze. How did Owibakum get to the quarters?" Jesse laughed.

"He had a lucky bye, a good win over the Filipino, what's-his-face?"

"Luis de Leon," Jesse informed him.

"Yeah, Luis. Then Mariano pulled one of his usual shitty moves. He beat the young hot-shot from Boston—Sullivan, Dan Sullivan—then walks straight to the referee's offices and scratches. Said he had a pinched nerve in his neck. I know for a fact that was bullshit 'cause I saw him practicing later at Roehampton when I went out to watch the qualifications. The least he could have done was to scratch before playing the kid. You know, give him a break—get through another round. That gave Owi a walkover. I think Sullivan would have beaten Owi. Then he would have given Shep some trouble. As it is, now you have to play Shep when he's fresh as a daisy."

"Oh Christ, I don't care—I'm fit. He's great to play. I always enjoy that." Jesse put his hand on Karen's, communicating to her his nervousness. Few people knew how tight he was before a match. His face never gave anything away.

Karen broke into conversation. "Did you know that was Missy playing the piano? The regular player just came back."

Wick pulled a face and waved his hands in disgust.

Jesse looked at him quizzically. "You'll have to tell me some day just what went wrong between you two. What did she do to make you so bitter anyway?"

"She's a horse's ass. I don't want to discuss her. How would you like it if you were making love to someone and they said, 'Don't come yet, I haven't thought of anyone.' Did Jaguar have an easy time with Freddie? I didn't hear the score."

Karen was used to their shop talk and casually watched the comings and goings at the bar. Missy had moved over from the piano to sit with Shep, who had just arrived.

"No, hell no! It was 7-5 in the third and the first was a tie-breaker. You play the way you did today and you'll wipe him up."

"Don't I wish. Wouldn't it be lovely if our best was our normal?

It's not like the old days, though. No, with just a few exceptions, anyone can beat anybody else on a good day. Sammy Kerwin, for instance—you remember the fat bloke at Queen's with Shep? You played him a—I think—quarter-finals or something long ago in the Orange Bowl."

"Can't be the same Kerwin, surely."

"Oh, yeah, and he's not such a fat-ass now and he's caught fire. Creaming all those qualifiers left, right, and center. Hey, with all due respects to the Danes, I can't stick this beer. Let's go to the Hansom and have some proper Foster's."

Jesse had started shaking his head, No, halfway through the question. "Karen doesn't even have to talk me out of it this time. Now I know you want to celebrate, but—Chroist, mate—you have a semi, I have a semi, and we have one together. Now don't go get pissed someplace. You'll need it all to take Jag. We could have an all-Australian finals."

"Yeah. You and I. Charming! You don't have to lecture me, Jesse. Look at Shep scouting around for extracurricular pussy." (Laurie Silverman had joined Shep and Missy). "Someday he'll OD on sex. I know, I mean I *know*, Jaguar will be at the Hansom."

Karen chirped up with, "Now there's a wonderful idea. You take Shep out and get him pissed—and Jaguar as well—then you all have smooth skating, just like once around the rink." Her beautiful skin was glowing with the orange, flickering light and VOs.

Wick started to get up in order to invite Shep—hesitated until Missy kissed Shep, shook hands good luck with Laurie, and exited—then launched himself across the lounge to the bar.

Jesse quickly signaled the waitress and paid the bill. "Wicko is still upset about Berconi and won't be happy until he wipes it out. We'll slip away before he gets back." Jesse held her chair.

She asked him as she stood up, "What finally did become of Berconi?"

"A doctor at Queen's gave him a tranquilizer. Now they have him under heavy sedation at Middlesex Hospital. He really did flip out. It wasn't an act."

"The poor fellow," she said, without much conviction. She didn't know Berconi and felt that anything that made her husband's life easier made hers a little more palatable also. Abstract empathy wasn't her bag.

Shep had back-combed his hair with a hand dryer and covered up his bald spot, looking his own age for a change rather than prematurely middle-aged. In shiny black silk slacks and full-sleeved white silk blouse, Laurie was striking. She carried her height with unique grace. Laurie knew how to walk and had a careless elegance. Her fine posture implied dignity—and she did have that.

As Shep wasn't making an elaborate play or turning on his charm full blast, she felt relaxed and held her drink as if she were doing a commercial. Wick waded up with his usual big grin. Shep said to him, in lieu of hello, "Wicko, you've got thirty-two teeth but they're all on top."

"You great old flatterer, you! Hi, Laurie. You look really nice—no, you do! Hey, Shepo, How's about you and Laurie joining us at the Hansom Cab for a cold chicken buffet and some good Australian beer?"

"Who is 'us'? I think you'll find that the Frasers have slipped off on you."

Wick looked over his shoulder. "He's been doing that my whole life, the great shit-ass. He'll have it off and be asleep by eleven o'clock. Maybe that's the way, but I couldn't stick it."

"Unfortunate turn of a phrase," Shep chuckled.

Laurie took Wick's arm and brought him closer to the bar. "We're going to Trader Vic's for some Chinese food. Why don't you join us?"

"Nah! Those great festooning, pufta drinks—and a half hour after I eat a Chinese girl I'm horny again."

"Whoops!" Shep censored. "A Chinese no-no. Look, I'm sure my doubles partner, Freddie over there, would be delighted to go with you."

"Oh—right! Moore's a good lad. He'll sink a few tubes with me. Good luck tomorrow, Shep. I hope Jesse knocks your ears off."

"Thank you for those kind words. I think."

"Fantastic win today, Al, really!"

"Thanks, Laurie. You and I have a lot in common, you know that?"

"Like what?"

"We're both the same height. Play well tomorrow, ya lovely thing. You're a sweet Sheila. Uh—don't wish me luck. I used it all up today. See ya." He sauntered off to talk to Freddie Moore.

Laurie asked, "Do you think we should have gone with him? He seems lonely."

"No." Shep fumbled with his money clip and dropped it on the floor. He continued speaking from below the bar stool. "He likes everyone but he's a little uncomfortable with Americans. He came from the bush and he's timid." He stood up again. "He'll be happier with the boys. All set?" They went out into the cool rain and queued up for a taxi.

The Hansom Cab was thronged—humming and throbbing with players, aficionados, and the hapless regulars who were testy over the invasion.

"The locals look disgruntled," Freddie said as he and Wick made their entrance.

"Come on, mate. Have you ever seen them look gruntled?" Wick replied as he elbowed a space at the bar.

"You know—I never did understand that word," Freddie considered. He frowned and pushed his lower lip out.

"Don't do that. Ya look like a flaming monkey." Freddie tried the face in the mirror. "You're right."

The cab was swinging from the ceiling from the sheer vibration of noise.

"Want a pint?"

"What?" The clamor almost reached the threshold of pain.

"A pint of Foster's?"

"Certainly!"

"Freddie, let's grab a plate of chicken and move 'round the other side. It's too fucking noisy and crowded here."

"What?"

"Shit! Come on." Holding their beers high, they pushed through the solid tangle of bodies. "Too many centers of the universe here tonight."

"What d'ya mean by that, Wicko?"

"Oh—it's just something Shep told me on a plane once. Here. This is good."

In the backroom side of the large bar, several tables were clustered where the royalty of tennis held court. Consuelo was sitting regally at a table she shared with a handsome South American type who looked like a 1930 matinee idol (moustache and shiny hair) and two other girls. Wick could make out the profile of Angie Redfield but didn't know the other.

Consuelo stood up to use the ladies' room.

"Look, Fred," Wick said. "She walks back on her heels and sends her snatch out as a scout."

"Well, I can see what you mean, but I think she's a great girl. I like her."

"I think she's all right too, but she still walks like her pussy was a torpedo tube."

Jaguar hollered over from a corner table, "Ho, Wick! Stop ogling Consuelo and come sing a song!" Jag was surrounded by the usual hangers-on. Billy was there too with a gorgeously stupid girl. "Still in her pink sweater," Wick thought. "We'll be over smartly, Jag. Hey, bartender. I'd like a split of champagne, a brandy Alexander, and a Harvey Wallbanger sent over to that corner table, my good man."

"Not bloody likely," the barman answered, surlily.

"Right. Well, in that case, just give us a couple of pints of Foster's."

With a grating of chairs and some audibles from scrimmage, room was made for them at the table. "Well, then. What would you like to hear? 'I'm Just a Bug on the Windshield of Life'? Or 'I've Got the All Overs for You All Over Me'?"

"How about the French one?" one of the white faces asked. Wick sailed into "Your Lips Tell Me No, No, But There's Wee Wee in Your Eyes," and they all joined in, applauding themselves wildly at the end. Consuelo was returning to her table and Wick was about to comment before a sharp elbow jab from Freddie silenced him. They were both privy to Jag's secret.

The group was discussing the uncertainty of Berconi's situation. The main thrust was whether or not the Players' Association would back him if the doctors pronounced him fit to play Wimbledon. Jaguar conjectured that the tournament officials would hardly permit the possibility of a replay of his streaking in front of the Queen. The group resolved the question to their own satisfaction by predicting that the doctors would tactically ask for more observation time, thus effectively keeping Berconi out of Wimbledon. Maybe.

"Jesus! Not another boycott!" Wick exclaimed. "The tennis public just wouldn't stand for it—even though we haven't had one for ten years—and I think the officials would be right not to risk it."

They all agreed. Someone contributed, "It's certainly going to give everyone in the seedings a jolly good boost."

"Yah! And seed someone else who's probably been grousing his head off."

Dirk broke in, "Do you think the women should get all the prize money they're screaming their heads off about?"

Jaguar was one of the last great machos. He had deceived himself into the belief that he was a fair and free thinker on the subject because women who wanted to be banged by his reputation and pale blue eyes fed his folly.

"I think," he began, "that they should get the same in tourneys that are two out of three sets. When it's the best of five, they should be ... you know ... whatever they call it ... to make it even." Someone pointed out that they were being paid commensurately with their skill and drawing power, not by the hour. Jaguar never deigned to look in the direction of the heretic but mentally marked him down and continued. "Men, most of the time, are more skilled and if they use those skills longer, like three out of five, they deserve more lolly and that's that."

The same voice piped up, "If Billy Jean King or Consuelo Alvarro had the speed and strength of a man, either one would probably beat you with their skills."

With a fine edge of pique giving a rasp to his voice, Jaguar tried to squelch the intruding element. "Maybe, maybe not. We'll never know, will we? But even so. How many Alvarros or Silvermans or Everts are there, eh? In women's tennis, anyone below the top five should take up archery or bowling. They're not in the same class."

Jaguar gave substance to the axiom that young Englishmen didn't really like women much. Oh, they liked to romp with them in bed, but when that appetite had been sated, they preferred the company of the lads, or fellows, depending on class. Then they could have their pint or glass of claret and discuss their conquests along with football and cricket. The upper-class men, being gentlemen, didn't actually discuss their good fortune with certain young ladies. They merely hinted broadly at it. For instance, "Lady Cynthia is a really ripping good blow job."

"Jag," Dirk asked, as loyal straight man, "what was Mariano's plan, scratching the other day?"

With the back of his hand, Jag removed some foam from his upper lip. "Maybe he was hurt or—eh—you know, pulled a muscle. But ya can't trust him. Probably he didn't want someone to find something out, ya know, in a real test. He's been hanging around with some bad types where he gambles near Knightsbridge. I'd report him to the Association if he hadn't had the boot already."

A languid, fairly attractive groupie asked in the hollow of the moment, "Jag, baby, will you play tomorrow with all this rain?"

"Oh, yeah, sure. They have four courts covered and, you know, play doesn't, uh, start till two o'clock. The wind and sun will dry it up on the others. A little slow and heavy, they'll be, for mixed doubles . . . but if it keeps raining, we'll go inside on the boards."

"But it's a grass court championship," the intruder said, about to leave.

"Well, they never play the *finals* on wood, you know," Jaguar said icily. "If it's too wet for the grass, the finalists would share the prize money."

The stranger was gone in the crowd.

"Who's your friend, Wick?"

"I don't know him. Only seen 'im around a bit." Someone offered that he was a reporter for *World Tennis,* based in Texas.

"Well that explains it. One of the Heldman Mafia. I . . ."

"Hey, Jag, whoa—Jesus! Give it a rest! You lose your breath you lose your turn around here. Bit of a bore, isn't it?"

"I don't know, Wicko—a bore is a person who talks about himself when *you* want to talk about *yourself.*"

Freddie heard himself saying into the short vacuum, "You know where the word 'bore' comes from, don't you? I was born in New Zealand and we have them." Many nods. "They are tidal waves that go up river just before monsoon season. In Burma and India and all. The English had their houses and compounds down by the river. Just before the 'bore' came, and they almost always knew when, they'd have to move lock, stock, and barrel—kids and servants, the lot—all up to these houses they had on the hill until it was over—and stay there for a week or more until the mud oozed back out. Then they'd say: 'What a bore! A crashing bore! Fucking awful bore!' " Freddie, pleased with himself, took a long pull on his beer.

"Fucking awful bore is right, Freddie," Wick said, guffawing. They all laughed loudly. "Very edifying, I'm sure, Freddie."

Maintaining the theme, Jaguar told them that the derivation of the word "posh" was a code of the rich English travelers on the old P/O Lines when reserving their cabins against the sun's heat. Posh were the initials for "Port Out, Starboard Home."

Sammy Kerwin leaned over from the bar and told them that jazz was originally a Negro slang word from New Orleans spelled *J-a-s-s.* "It means 'fuck.' 'The Jass Me Blues.' Basic, funky jazz music. Sounds like what it means, brother."

That was a good one and they all privately registered it in their compendium. After several rounds of beer and dirty jokes, the bell for time rang. Wick threw back the last of his drink and said, "One last. Do you know why pubic hair is curly? 'Cause if it were straight, it would poke your eyes out!" The laughter wasn't as resounding as he expected so he added, "Um—well—it's just a little joke, and when they're that little they need a lot of love." He found his feet among their smiles.

"What about gruntled?" Freddie said, with an aborted belch.

"Sure, Fred—let's go home while we're all still gruntled. Sleep tight, all. I will." They sang an Australian song together out into the London mist.

21

The luminosity of the white bahias in their long, tulip-shaped glasses made them seem like beacons amid the various plates of Polynesian food. In their initial hunger, they had ordered all the traditional Chinese hors d'oeuvres, from egg roll to butterfly shrimp, with all the sweet and sour betwixt, *and* dinner. Now there wasn't an inch on the table that wasn't covered by food. Even the ashtrays had been removed to make room—not that either of them smoked anyway.

"Mother used to say to always have a sandwich before going to the supermarket and you would only buy half as much as if you went there hungry," Shep said. "We should have stopped for a Wimpy Burger. Now we wouldn't be intimidated by all this stuff. We really *are* prisoners in a fortune cookie factory."

The waitresses arrived with another course, and they both gasped "No!" in pain.

"I was brought up with a sensitive conscience about not eating my food because of all the starving children in India. We have enough for two families here and I'm really ashamed," Laurie said with a groan. "These slacks were tight when I put them on. Now I'll be afraid to bend at all."

She was the proud owner of a behind of classical line. Wonderful impassioning slopes and curves. Seeing her in butt-sprung tennis shorts, Shep had never guessed at the disguised treasure. One thing he had been aware of, though, and he was admiring them right at that instant, were her dynamite boobs. They started right from the upper chest and crowded out against the blouse, the nipples going erect from time to time inside the silk. He would like to have thought that he was the cause of the concupiscence, but knew it was probably the rubbing against the silk.

The bahias looked and tasted like milk shakes to Laurie, but they included two types of rum and she wasn't accustomed to alcohol.

"Shep, I can't eat another chopstick full of anything and I'm slightly gaga. What time is it getting to be?"

The staggering bulk of food, the circuslike day, the rum, and intermittent lust were taking the toll of Shep too. He felt that the tikis, outriggers, and walls were closing in, and he was in a heavy, smiling, semi-stupor of contentedness. Even the sound of their voices had become hypnotic. The evening and conversation had gone smoothly, with a rich range of topics from Mendelssohn to baseball, politics to Braque. Laurie had taken her degree in art history and was very pleased to find that Shep was more than sophomorically knowledgeable. The year he had torn his hamstring and couldn't play, he returned to school to get his masters in psychology. He enrolled in a crash art course that opened up the aesthetics of life to him, not merely man-made, but the full spectrum. He was not just "into" art but aware of the constant beauty of nature and the wondrous order of things.

They'd had a great old time together developing ideas that they had forgotten they'd owned, and the communion of thought and mutual taste made them very close.

"Shep, you'll have to wheel me out of here. I'm in a trance."

"We'll be like Tweedledum and Tweedledee, then, because I'm not exactly fighting fit either. Shall we take a dragon bag with some of this stuff?"

"God, leave it!" With a flash of daring, they finished off the bahias and gathered themselves together. Shep signed the credit card slip and they floated lightly up the stairs.

The rain had stopped. Billowy, dark clouds swirled across patches of stars. Shep wondered who had first coined the term "scudding" for clouds. It was a perfect description. He said the word aloud.

"Beg your pardon?" Laurie said.

"The clouds—they're scudding."

"Oh, are they doing that again?" They floundered inside a cab and headed for South Kensington.

Shep was a bit chagrined when he realized that only the first- and second-round losers were still up at that hour. Jesse, to be sure, was long asleep. They went up the lift to the same floor. His room was just down the hall. He waited until Laurie put the key in the lock, then turned her gently for the traditional first-date good-night kiss. It started out well intentioned, but it lasted, lingered, became heavier and more rhythmical. They one-stepped inside, out of the corridor, around the door, and Shep elbowed it closed without interrupting. Laurie felt a blazing flush, and the rum throbbed along with the blood in her temples. Very, very gently, Shep cupped her hot breast and stroked her jutting nipple with his thumb. The top button slipped open by itself. He worked his hand inside and sculpted the other one out of the blouse; blue-veined and satiny white below the tan-line. The nipple was very round, very erect. He inclined his head and took it into his mouth, sucking, biting gently and brushing it with staccato motions of his tongue.

The sight of her own naked breast in his mouth excited her to hot breathlessness and she fumbled with his zipper. Shep released her clasp and, unzipping her slacks, slipped his hands inside and behind to grip her voluptuous ass. The slacks slithered down to her ankles and she kicked them off and across the floor.

"Wait a minute," he whispered. "Let's take off our clothes and do it properly." He knew she wasn't going to change her mind. He could always tell.

Laurie knew it was undoubtedly the rum, but she was literally on fire, with that hot ball of passion burning in her throat and chest. Two blouse buttons and a swimming motion and Laurie was naked on the bed. She was a long, Levantine "Maya."

Shep was memorizing her. He turned on the small reading lamp, not wanting to miss anything in the dark. Even though she was lubricating wildly, Shep had a difficult time entering her. After a few minutes, he allowed her to roll over on top. By phases she lay, sat, and finally squatted on him, raising up and down slowly, staring him hotly in the eyes. He held both her breasts and manipulated them alternately clockwise and counterclockwise over his face. Then she leaned back from him and inched down the bed on her knees. She enveloped his prick with her pliant breasts, and then, stretching out, took him in her mouth.

Shep threw away the mental manual on technique and just allowed himself the animal ecstasy of the sensation. When she judged him about to come, she sat up, spun around, and sat on it, and, with three fingers, massaged her clitoris with gentle jabs. It was the stars and fireworks and all the clichés, real, valid, and it was *their* moment—and they were very alive in it. Their entire bodies, every cell was coming. Billions of minute pleasures—and Laurie was moaning loudly, "Put that fuck into me, Shep—owwww! you lovely bahia!"

After several quivering minutes, she arched her back, released him, and turned to snuggle up beside him. There was a chilling knock on the door. "Laurie, are you all right?" They both went up on their elbows, heads forward. It was the Swedish player next door.

"Yes, thank you, Inge. It was the television. Too loud. Good night." She giggled nervously and wrapped herself up in him, their legs entwined. It was wonderful to have someone longer than she to cuddle. She felt less like an Amazon. "I thought it was my mother, for Christ's sake! Whew!"

"Aren't you a big girl now?"

"You know better than that. When is a daughter ever grown up to her mother?"

"Yes, I know you're right." He nibbled her ear. "I can't stay. We have to get our sleep."

"You can't stay anyway. I feel simply wonderful."

"Laurie, that was only magnificent."

"Hmmm—yeah. Someone once taught me those variations on a theme, but it's the first time I ever enjoyed it. It always seemed so decadent and sort of like a lifeboat drill or something. I'm afraid I'm a fallen woman. And you're only the third man I've gone all the way with."

"Do people still say that?"

"This one does." She kissed him long and lavishly, using her full lips hungrily. "That was good night—for sure."

Shep swung his legs out of the bed. "I guess I'll have to back out of your room with my hand out."

"Great idea! Do it." Laurie got behind the door when he was dressed. He opened it and backed out, saying, "Good night—lovely evening."

She was humming, "Bayee—by-ee-i-o, Bahia . . ."

The dawn didn't look like dawn. The horizon was simply less grey than the rest of the sky. The clouds had knit solid through the night and now massed low over London. Shep pulled the curtain back, his mood going from broken-sleep cranky to schoolboy glad that he could jump back in bed for a few more hours. He thought of Laurie and drifted into sleep with a smile.

Across town, in a rented apartment in the West End, Karen squinted at her watch and decided to let Jesse sleep a little longer. His warm, sleeping body next to hers was reassuring in its alpha-calm and she savored the moment, vividly aware of being extremely happy in the little captured time capsule of their bed.

Instead of taking "Miss Languid" home from the pub as all of his better instincts screamed at him, Jaguar drove out into the country to stay with Consuelo, even though he knew it would be a Latin, pre-nuptial night. She slept virginally under the sheet, and he on top on the outside. The Oxfordshire country night was black and heavy with peace. Jaguar reached out and touched Consuelo. Even far into sleep, he desired her. She scratched the spot he'd touched and, murmuring, curled up tighter with a baby pout . . . both pressing deeper into sleep.

Wick's own snoring startled him awake with a short jerk Shaw's line popped into his head (Wick was a sneak reader): "Laugh and the world laughs with you, snore and you sleep alone." The tiniest suggestion of a smile tickled the corners of his mouth as he thought of his handiwork, a prank he had set into operation earlier in the night. A potential laugh for the future. Right now he needed sleep. Big day ahead. He lay still, listening to the telltale squeaking of bed springs somewhere in the flat. "Shit!" He turned over, pulling the pillow tightly around his head.

Eduarde Vector drank little and infrequently. He thought gambling a waste of time and rarely smoked—and never pot. Eduarde Vector had one vice and it affected him, when he had an attack of it, as though his blood were replaced by chili sauce. Women—fat women —the fatter the better. His minimum was a normal man's maximum. Even now, in his sixties, the urge stampeded through his hot checkpoints from time to time with a fierce urgency.

Vector had another kink. Hoarding tennis balls. He bought them from tournaments at a third off as once used. (They usually had only nine games on them. In major tournaments, the balls are refrigerated and changed every nine games.) Eduarde Vector had enough money now, but the old habits die hard. He was going to take them back to Australia and sell what he didn't need for his own use—and he had thousands of them, all in huge cardboard cartons ready for shipment.

In the little hotel Vector had chosen while in London, there was a stupid but strict rule against women in the rooms. Not that there was anything special about the rooms: shabby dresser, Victorian wardrobe, overhead light, and blank walls. But it served Vector's frugal tastes and purpose just fine otherwise. Except for the rule about women.

Eduarde Vector had met one that very day. "A fifteen stoner," by Wick's estimate. She was well over two hundred pounds. Melanie Putzer had smiled with a simple-sweet face at Wick when Vector had introduced them that afternoon. The hideousness of the idea had made Wick shudder with glee and horror even as it occurred to him.

At closing time, Wick and Freddie Moore left the pub and went to Vector's hotel, using the same route he would use to bring the lady upstairs—the fire escape to the second-floor landing. They dumped all the balls on the floor, wall to wall balls a foot deep. Then, on Freddie's shoulders, Wick had reached up and unscrewed the light bulb.

Eduarde Vector and his fat mama crept up the fire escape and along the dark hall to the room. Vector turned the knob and pushed. It was stuck. He shoved harder, opening it an inch.

"Gif me a little hand here please, Melanie," he whispered. Together they rammed the door. It crested a wave of balls, and they fell inside in a huge heap, floundering about in the dark sea of balls, stumbling, sliding, up and down, she screaming, sure that she was in the clutches of some kind of sex pervert. (Tennis ball freak, she would have called him if she had known what they were.) Eduarde Vector was swinging his arms about, trying to find the string to the light. He felt it slip across his face and grasped it. Two tugs produced nothing. On the third try, he lost his balance, rolled belly-up in the air, and trampolined down into a blackness of fuzzy rubber. The string came with him.

Somehow, fat Melanie made her way on hands and knees to the door, yelling all the way in banshee blood curdlers. She half stumbled, half rolled into the hall and sat on the floor with her dress up around her waist with two yellow balls between her legs. She looked like the goose who laid the golden eggs. The hall lights blazed on. She blinked at the night clerk, who demanded, "Here, what's all this, then?" A waterfall of balls bounded down the stairs.

A kindly constable talked the manager into allowing Eduarde Vector to finish the night (without Melanie) so that he could repack all his balls before being evicted. There was no doubt in Vector's mind who the guilty party was, and he seethed as he lay quietly among the motionless mounds of medicine-smelling balls. It began to brighten a little outside as he fell asleep, a slight glow of yellowness around him.

At six in the morning, with a semifinal doubles match to play that afternoon, Randy Mariano was still trying to get even on roulette. Any hope of a big coup and breaking out of his bind had disappeared long ago, about false-dawn time. With a reasonably good run (considering the way his luck had been going in the past weeks), he had taken two thousand pounds from the blackjack table at the Brass Shoe Club. Relatively speaking, it amounted to nothing, since he owed well over two hundred thousand dollars. More, actually, since a check he'd given that night for partial payment and chips would ricochet across the ocean. He wondered how the casinos at the turn of the century managed when it took two months for a check to bounce, but realized that they probably only accepted letters of credit from anything less than a billionaire. Then they would, of course, allow one to play only up to the limit of the credit.

It had been a disastrous year for Randy. Nothing bloomed. He was a good, cool, percentage gambler, but his luck, the quintessential ingredient for even the most skilled player, had shown out badly time after time in the clutch.

The night had started with fabulous promise. As he'd wandered among the tables, the mocking glitter of the chandeliers, and the noiseless, smoke-filled red carpets that didn't even allow a man a footstep, good omens were everywhere—private little bits of good luck in the offing which he alone recognized. The fact that they'd cashed his check was indication enough that this would be the turning point, even though the major portion of the cash would have to go as partial payment of the National Debt, as he'd come to call it. England or not, these people were not to be fooled with. Few Americans realized the capacity for brutality the British gangsters had . . . which seemed even more appalling in a country famous for gentility and fair play—you win, they pay; you lose, you pay—and he'd dropped the hairy end of the lollipop that was his end of the contract. And as gambling debts aren't legally collectable, what's a girl to do . . . ?

He had made two good hits on full numbers at roulette, still playing with house money he'd won at the twenty-one table. It was *all* house money, but they didn't know that. At that moment, somewhere around three o'clock, he had run it into $23,000. Enough to redeem the check and still have $3,000 to play with astutely and prudently. Then, at least, he would have paid the interest on his debt up to date and shown good faith.

With a flying sortie over to the Playboy Club by cab, he permitted the myth of his luck to more deeply engulf him in the bottomless abyss of his illness. He won another thousand just walking around from table to table, but was unimpressed by the types there that night (they didn't recognize him as the great American tennis player he was; even in debt, he was a somebody at the Brass Shoe) and he squirted back out into the early damp. At the Claremont Club, the door ensemble cordially shook their collective heads "no" at his arrival. With good humor, still riding on "his night," Randy said a cavalier good night to them and popped into Annabel's for a small drink and look around. Of the tennis crowd, only Angie Redfield, with a girl in a pink sweater, were there; actually passing him on the stairs on their way home.

After a professional glance around, he thought it better not to mix alcohol with the pep pills that were helping him to make it through the night and grabbed a taxi back to the Brass Shoe.

Somewhere around five o'clock, his luck, most of his money, and a blond shill who had become his charm all deserted him. One hour later, he put his last stack of chips totaling a thousand dollars on number 17, black—the table had thinned out, but there were still heavy rollers placing the chips in recondite places, their dead eyes giving no secrets away. The wheel spun—it seemed on this roll that nothing on earth could stop its spinning. Finally it slowed and the little ball pitter-patted frivolously around the numbers with its malicious, nasty jauntiness and rested on number 4. Randy had the satisfaction, fleeting as it was, of noting that no one else was on the number either. He walked off, out into the dawn. At least the goddamn sun wasn't shining.

22

Shep's semifinal match with Jesse, the first match of the day on court one, would have been a final in most tournaments of the world, but Wimbledon was the "gathering of eagles." From Davis Cup matches, from International Team Tennis whose contracts burdened the players with commitments right up to the eleventh hour), from other warm-up tourneys around England and a minor circuit in Africa that competed innocuously with the other circuits (making do with a few name players in the dusk of their playing fame), they continued to arrive. Very few were prepared for the grass and were grasping at any possibility to accustom themselves to it in three days.

The African circuit offered a more viable potential for the production of good, perhaps great, black players than the American did. As Arthur Ashe and black sociologists had pointed out, the best

American black athletes were going into team sports, not tennis. And it wasn't just a matter of money. The draw of the tennis money was now incredibly tasty, but in black America, even with street tennis and the allotment of a few hours a week of school gym time for tennis, whenever ten youngsters could use the only spacious area in a crowded, competitive, city situation, basketball would be the game, not tennis. Therefore, it was not surprising that the two best black players in the world had just arrived not from the States but from Africa and were shaking hands with the boys in the dressing room. Either one could be very dangerous on any given day. Both Jesse and Shep respected their racquet work.

"Great to see you again swaggies," Jesse welcomed them. "I wish you luck getting some practice in on grass. In the meantime, get on the wood. It's the closest thing."

Shep called to Jesse, "They've announced us."

They walked downstairs together in silence. Both were deep in concentration over strategy. A standing ovation brightened the dark day as they walked on court.

"Rough or smooth?" Jesse asked.

"Smooth."

"Rough it is. I'll serve." Jesse sat down to make an adjustment to his shoe. "Shep . . . don't look now . . just over my shoulder scouting us. Mariano. His eyes look like flying saucers."

"Frightening looking. You're playing him in the doubles later. Get him out of here altogether. He gives me the creeps."

Once again the silence clanged closed between them and they maintained it. Neither of them subscribed to the theory that one should hop himself up on hating his opponent. They had discovered long ago that this type of psyching was, at the very best, counterproductive. They knew each other's game and were dead even in the won-lost column. Winning now would be a very fine psychological advantage, coming so close before Wimbledon. They both settled down mentally to give it a full shot. If Shep's theory about sex was valid, they were equally in trouble.

After a nervous start—a double fault and a volley over the baseline—Jesse shook most of his tightness and serve-volleyed sharply for the first game. Shep, angry for not having capitalized on an early break-of-service chance, started shakily himself and was at 30-40 before his volleys found length. Overhead, he was very solid, and the two tentative lobs Jesse tossed up to test were thumped away with pistol-shot resonance. The tennis was workmanlike if not brilliant, full of cunning and technique, although it looked slam bang to the audience.

Between games, with Shep behind 3-4, Phil Katz tried to hand him a telegram.

"It's from California," Katz said.

"Hi, Phil. Yes, my dad's coming in sometime tonight or tomorrow."

"It's marked *urgent*," Phil stressed, feeling awkward and out of place on the court.

"I'll read it between sets," Shep said. "Excuse me if I try to save this match?"

Shep grabbed his racquet and hurried out to serve. Jesse was already out, pawing with his foot at a rough spot of turf behind the line. The court was holding up well thanks to the overtime sweat of the grounds keepers. The canvas had kept it completely dry. Behind the baseline, on the four service areas (two farther over for doubles), there were bald patches. There was another just behind the "T" of the service line where the players pulled up for their first volley and two more small areas of earth on either side of the centerline closer in, where the second volley was hit. It was a graphic diagram for an intermediate player on exactly where to move for serve and volley.

Shep pulled his concentration in and around him and, putting Katz out of his mind, played airtight tennis. There were more rallies now that they were returning better, and flashes of the brilliance they both possessed began to be manifest, slowly at first, then with wonderful saves and touch.

On the change at 4-5, Shep again spotted Mariano among the faces along the first tier. He remembered something that Wick had told him and a huge grin spread sweetly across his normally dead-serious match face. Wick had said that when Mariano was walking onto the court with young Dan Sullivan, he tried to psyche him saying, "I know you're probably nervous playing me, but the supporter is supposed to go on the inside." The Irish kid shot right back, "How would you know? You don't even need one!"

Jesse's eyebrows arched wonderingly at Shep's smile, and the gallery hummed in curiosity over his secret. Head to head they battled, giving nothing, running down, up, and around every ball that still had movement. Touch and power, cross-courts and short, drawing balls—on and on until at 8-all, with no service breaks, they went into the tie-breaker. Lingering death, as the longer European tie-breaker was called because it had to be won by two points and there was no logical reason why it need ever end. Phil Katz had waved the telegram a few times during the subsequent changeover, but was now sitting quietly with a stone face. Shep didn't even glance his way.

Shep jumped away to a quick 3-0 lead (needing 7 to win), winning Jesse's first serve on a passing shot and serving two unreturnable balls on his own serve. Jesse got back into it with two smoking serves, one which Shep only just saved from being an ace by a touch. The second was sent feebly but dangerously short over the net. Jesse glided in and, with a bolo swing, imparted driving topspin to it, whipping it up the middle; no percentage, but demoralizing.

The point that brought it to 3-3 was the shot of the match. Shep served, medium-paced and well placed, wide to the backhand. Jesse stepped well across and chipped it heavily cross-court. Running flat out, turning sideways at impact, Shep hit a topspin backhand down the line that looked a winner. Anticipating the placement, Jesse had already jumped off to his right, and he saved it with a defensive lob well over Shep's killer smash. His momentum had carried him right up to the net. Jesse, in the meantime, had slid on the grass across what Wick called the "taint," between the court and the stands, and ended on his ass with both feet under the short side canvas. Shep, running back, never saw him fall down. He had his own troubles. Back on his feet, wiping his hand on his shorts as he ran, Jesse took the net—only to have Shep hit a perfect lob over his head. Seemingly perfect. At full run, leaping and twisting, Jesse hit a crushing smash just in front of the back canvas, for a winner down the center of the court. It dug up a divot of sod where it struck. The applause for both of them lasted over a minute. A minute they both needed.

The long point, plus the changing of courts in the middle of his service, upset Shep's concentration for the moment and he served his first serve just long. Jesse pounced on the unusually short second serve, powdering a cross-court backhand winner—his bread-and-butter stroke. It gave him a 4-3 lead in points and the edge in momentum and confidence. He had won four straight points. But Jesse made an error in tactics deciding to swing a first serve down the middle (in the ad court) hoping for a surprise winner; Shep hit a clean, clear-sounding shot so hard it was a blur as it went by Jesse. Shep put a return of serve out, and Jesse led 5-4 with Shep then serving two. A bad bounce off a chipped return gave Jesse 6-4, but Shep banged down his seventh ace of the set for 6-5. The damage had been done, however. Jesse came in with an awkwardly bounding first serve that Shep mistimed. It struck the net-tape, teetered on the rim, then dropped back into Shep's court for 7-5 and the first set.

They started to change courts for Shep to serve the beginning of the second set when Phil Katz stepped in front of Shep.

"I took the liberty of opening the cable, Shep. You'd better read it, fella."

Shep opened the folded paper:

FATHER HAD MASSIVE CORONARY NOT EXPECTED TO LIVE THROUGH NIGHT COME HOME INSTANTLY I NEED YOU

MOTHER

Shep looked at the ground and emptily, with a voice barely audible, said, "Oh." He handed the cable to the umpire and said, "I have to default." To Jesse, who had come back to the chair, he said, his eyes filled with tears, "Sorry." Leaving his gear for Phil to collect, he made his way blindly up the stairs to the changing room through the silent crowd.

The umpire started to make the announcement. "Ladies and gentlemen, due to a . . . uh . . . personal tragedy, Mr. Shepard has defaulted. Fraser wins, 8-7, resigned."

Jesse vacillated for a full minute or so, arms at his sides, feeling quite useless. Any gesture or word at that moment would have been futile, he knew, but goddamnit anyway. Methodically, he picked up his things and cleared off amid—nothing—the gallery couldn't applaud, express sorrow—its presence made apparent only by the low rustle of whispers. He didn't go straight upstairs, reluctant to intrude on Shep's sorrow by having to commiserate. Instead, he went into the tearoom to try to wash down the lump in his throat.

"Bloody shame, Jess," someone called out, covering everything.

"Yeah," he mumbled.

Wick stopped in on his way on court. He put his hand on Jesse's shoulder, saying, "Apart from the main sadness of it, it's a pity the game had to stop. That was some of the best tennis I've seen since I watched Rosewall and Richey at Wimbledon when I was a kid."

"Ah well, swaggie—we'll play again next week. Ya know, 'the dogs bark but the caravan moves on.'"

Wick had a distant expression in his eyes as he remarked, "I wonder what Gurdjieff would have to say about that?"

"What? Look, Wicko. Go out and play another great one. I'll shower and watch a bit. We've got the doubles later, but put that out of your mind."

"Right, mate."

The second match was a dismal shambles compared to the quality of play Wick and Jaguar were capable of. Wick couldn't find the range and was playing so badly he was making Jaguar look rotten in the bargain. So much so that Wick actually took the first set 7-5 with a combination of mis-hits, various other forms of luck, some occasional good serving, and a tentative Jaguar, who kept waiting for Wick to raise his standard—play the brand of tennis that had brought him into the semis.

But Wick only got worse. Still trying to find the corners, he couldn't find the court. Jaguar relaxed, played looser. He knew now that the sleeping lion wasn't going to spring to life with any surprises. Jaguar changed tactics. Instead of going for the big shot which he'd felt incumbent on him (after the dazzlers the audience had become used to), he simply hit three-quarter pace, medium-length shots and let Wick do the damage to himself. Jesse left midway in the first set. He was very well acquainted with Wick's off days and was ninety per cent sure of the outcome no matter who took the first set. What the outcome of their doubles was going to be he shuddered to think.

In the changing room, Mariano, now dressed, looked like a mutant out of "Star Trek," his eyes saucerlike on whatever pills he was popping. Mariano's partner was Mikhail Pakachev, the Russian number one. Pakachev, after having a row with his Federation about more pocket money (they were giving him one hundred dollars a week after expenses out of his $50,000-a-year earnings), had disappeared from the circuit behind a curtain of secrecy. He had just as mysteriously surfaced a day or two before the Queen's Club tournament, and had found to his distaste that Mariano was the only quality player without a partner. Jesse hoped that Pakachev would make some sense out of their doubles.

Wick's Chinese firedrill, although three sets, was mercifully over in little over an hour. Wick, grim-faced and shaking, had gone up to shower and put on some clean gear for his doubles match.

Jesse went to the club lounge and found it clustered with people passing the time playing chess, backgammon, cards, or quietly reading. With some exceptions, they broke down into cliques—Russians, Americans, Aussies, and English playing the games; the Latins in groups talking rapid fire, always enthusiastic about something. Jesse allowed a huge leather chair to surround and protect him. His shyness safely provided for, he opened a *Daily Telegraph* and looked at it, not really focusing on anything. He read a whole column without comprehending a word. His thoughts were in flux. Unconsciously, he had identified with Shep's loss. Shep had the type of father he would have chosen. A bit similar to Wick's but with more grit. Jesse sat there comparing his own father to Shep's, or rather, to his ideal; contrasing the shame of his own father's bizarre end with—what? What did it matter how a man ended? Death was so irrevocable, dignity didn't matter to the dead person. Or did it? Wick was always reading preposterous books and had once read to him from, of all people, Leonardo da Vinci: "To think that all these years I believed I had been learning how to live, when in reality, I have only been learning how to die."

"Quit mooning there, Jesse. Let's get it over with."

"Thinking about Shep and things, Wicko. Between you and Shep and your lightweight philosophies, I could be round the bend. Shep once said—maybe this isn't exactly it—but: 'Life is a sea filled with people swimming with different styles, strokes, in no particular direction, for there is no shore.' "

"Bullshit!"

"Why do you say that, Wicko?"

"Because it doesn't conform to my way of thinking. So it must be, you see, bullshit."

"I see. That's a refreshing attitude. Very broad-minded."

Wick began rotating his head slowly, rubbing the back of his neck. "Some bad day I'm on to. I woke up with a jerk and got this sore neck."

"Depending on the jerk, you're lucky that's the only sore thing you got."

A bit short on humor, Wick answered, as they walked down the steps, "Don't make jokes, Jess. It doesn't become you."

Jesse waved to the referee that they were ready. "You're in a great mood. Did you get pissed last night?"

"Nah! Not really. I know when I'm really drunk 'cause I set fire to myself and start to giggle."

"That froggy voice of yours—you must have been doing your Sinatra routine, or flapping your mouth again."

"Just leave me alone. I just lost a semifinal, remember? Here comes that American reporter again. He's a right pisser, he is. Hey! Who're you looking for now?" The reporter was tall and gaunt— sharply pressed searsucker trousers, lightweight pale blue blazer with at least nine different badges and forms of identification, not to mention cameras dangling. His thick glasses made his eyes look like fish swimming beneath his receding crew cut.

"Hi, Al! What's happening? Yes, I'm looking for Randy Mariano —you know—like from California? Have you any idea where I could find him?"

Wick smiled, knowing that Mariano was standing out on court one—right behind him. "I saw him a while ago in the lounge," he said, "but he went out to take a shit and the hogs ate 'im." Jesse laughed and pulled Wick along before he warmed to the nonsense.

"Here come Batman and Robin," Wick said as their opponents walked out on the court. The ball girls tossed them balls and curls. Mariano towered over the short, swarthy Russian. No one would ever believe the miscasting; the Russian was the good guy and the American the bad guy. Jesse said as a throwaway, "Now, Mariano really *is* one bloke who thinks he's the reason he's so tall."

The applause was scattered. The numbers had thinned out but not appreciably; the after-work fans were filling the gaps. Jesse won the toss and took serve. Close up, Mariano's eyes looked like a pair of dry wells. Suddenly, for the first time that day, a perfect circle was sucked through the clouds like an inverted mushroom and the sun literally burst through. At five-thirty in the afternoon the dawn finally happened. People were blinking and shaking their eyes, a bit provoked by the latecomer. A nice warm aura cupped the court, and the players took their positions. Just as dramatically and rapidly, the sun disappeared for the day into the black nimbus; as if a light switch had been clicked, it was again dark—even darker.

At 3-all, everyone holding service, it was very apparent that Jesse and Wick were in trouble. Pills, lack of sleep, worries—nothing slowed Randy down. He was playing superbly. He and Pakachev had both won their serves with ease, while Wick was slopping about, just barely winning his from 0-40, and Jesse, uninspired, winning his after 30-all. There was no question in anyone's mind who was going to be the first to crack. This time, Wick lost his service. The Russian went out after the changeover leading 4-3.

Jesse provided several points for ad—it went to five deuces—but Wick couldn't capitalize on them, and Mikhail eventually brought his team to 5-3. The bar was now open and the noise from the club terrace rolled like thunder. Jesse won his serve and Mariano closed out the set 6-4.

"Get your first serve in, Wick," Jesse said, in as soothing a manner as he could without being patronizing.

"Actually, I'm throwing it away on purpose," Wick snapped, " 'cause my second serve needs so much work." He was clearly nettled.

Wick lost his serve to start the second set. Randy chipped beautifully at his feet as he came in. Up the middle to his backhand, but just out of the reach of Jesse's backhand poach. On the changeover, Jesse and Wick lingered halfway to the chair for a little conference.

"You're playing great, Jess. If you can carry me a little longer, I'll snap out of this and get into the match." Wick said this apologetically.

"I'm not playing that great, but in the valley of the blind the one-eyed man is king. We have to do something quickly—right now —or we can pack it in."

"I'm waiting for a quick rush of shit to the head."

"Wicko, don't try to think today. Leave it to me. Go down the line on the first chance. I'll volley his volley. Then we'll put up a hundred lobs."

Jesse didn't have to volley anything. Wick, stepping to his right, around the backhand, cracked a clean, clear whistler down the line. It sounded like a toy electric train going by Randy. On the next serve, Jesse lobbed high in the air back of the service line. Both he and Wick reverse-pedaled into the backcourt. Mikhail didn't hit a particularly good overhead and Wick sent it high to their backcourt once again. Randy let it bounce and smashed down the center. It was a good shot, but Jesse made a lucky half-volley lob that dismayed all of them. Wick and Pakachev played a little psychological pas de deux, Wick making feints to run in, Pak swinging into an artificial wind-up—that didn't fool Wick. When Pakachev made a holding motion, then just tapped the ball over, Wick was already on his way in and, sliding into position, he pushed the ball down the line for the point. Randy glared at Pakachev and began muttering *sotto voce* and shaking his head. Pakachev gave him the palms-up chin gesture.

A rapid-fire volley exchange among the four ended in the point for Jesse. At 15-40, Jesse again lobbed one with topspin to Randy, who tried an angled placement instead of a big shot. Jesse ran cross-court and belted a forehand winner up the middle for the break.

Jesse won his serve with the help of a good poach and smash from his net man, who was surely showing signs of life. When they changed courts, Randy made a snotty remark about Jesse's lucky pick-up.

"Yeah, Randy. Once again, science and technology prevail over ignorance and superstition," Jesse smiled.

"You're out of your tree, man."

"Like Wick always says, I like it but I'm not proud of it."

As they walked back out, Wick said, "Jess, don't waste your breath talking to that shit."

"I donno. He is a human being and he's looking worse and worse. Anyway, never mind that. Now listen to me. We're going into formation against his cross-court chip shot. He can't do that down the line. Remember on your serve. We won't have time to talk about it —I'll just do it."

On Mariano's service they lobbed, and lobbed brilliantly, from all parts of the backcourt. Randy and Mikhail made some and missed some, but the exertion of the smashing was telling on Randy and his pills were wearing off. He double-faulted for game.

With Wick serving at 3-1, they switched into the Australian scissors—both players on the same side of the court when serving to the backhand, the ad court. That way, Jesse cut off Randy's great cross-court clip shot and Wick went to net up the right side of the court rather than the left. The maneuver completely destroyed their opponents' rhythm. Wick won his serve at love. With the score 4-1, Randy had had it and the rest of the match became a formality.

Mikhail Pakachev looked stoical about it. He was only in it for the practice on grass anyway, and the now dissolved possibility of a win would only have been a little dividend—if Russians were allowed the luxury of dividends.

Even in the last flashes of his "uppers," Randy was still ranting along. "Pulling that formation was out of sight. Do you think that's what did it to us?"

Laconically, Jesse answered, savoring it completely, "Winners don't have to explain."

"Heavy! He-e-a-vy! So man, have you—like—got any tips for a poor hacker like me?"

"Yeah—stay away from the tables and try to find God."

"Right on! God's far out. What about you, hot shot?" he said to Wick, who was patently ignoring him as he walked ahead. "You're pretty quiet, man."

"Right. I'm saving my personality for tonight."

"You're dynamite, Mr. Wicko. I love you. So long." Randy went to his locker, still talking in the distance, "If I'd had a partner we'd have routined you, with Mr. Wicko there playing like a water buffalo in labor and . . ."

"You'd better not let Mikhail hear you talk like that or comes the revolution, you'll be sucking Gulag goulash," Wick shot back. "He looks like a tough little fucker as well. You'd better belt up."

"Better not let Mikhail hear what?" The Russian came out of the shower; his heavily muscled, stocky body looked more like that of a gymnast. Both Jesse and Wick turned around toward their respective lockers and simultaneously put a foot up on the bench like chorus girls.

"You are talking about my tennis?"

Randy tried to look off-handed. He was bored by the whole thing by now. "Sure—you *must* realize you played like four pounds of shit in a one-pound bag out there."

There was a moment, a clear hiatus when it seemed that Mikhail was waiting for the translation by earphone, then Robin fired off a left that caught Batman flush on the right side of his jaw and dropped him in the aisle. Randy didn't get up, indulging in his first nap in thirty-six hours.

Billy Sherman rubber-necked around the corner and immediately went for a towel of ice for Randy's face.

"You *are* naughty, Mikhail me lad," Wick said.

Jesse shrugged. "I don't think it warranted that type of action—but then I didn't have to play on the same side of the net with him. I reckon I'd be dishonest if I didn't admit that I'm delighted."

"I think my hand is broke," the Russian said.

Billy came back with the ice. "The fans and papers are getting their money's worth *this* week, what with Berconi's streak and the ladies' blow-up, Shep's thing, and now this. What a week. Last night Angie Redface scooped my date. Imagine being knocked out of the box by a dyke!" He held the ice against Randy's jaw.

"Who did Angie grab? Not the dolly in the pink sweater?" Billy nodded in reply to Wick's question. "Oh, you're well out of that, mate. That sweater must be bloody high by now."

"Maybe she has twenty of them like Little Orphan Annie. Anyway, I always thought I could cope with the situation if it ever came up. But what does one do? I couldn't very well ask Angie outside and say 'Leave my girl alone or up your nose with a rubber hose.' I just sat there with my thumb in my ass when they sweetly said good night—and had the cheek to kiss me on the forehead—that was the unkindest cunt of all."

Jesse snickered and shook his head in disbelief, very glad to be out of all that. Randy was sitting up now, still groggy. Wick addressed Pakachev. "I'll take you over to Middlesex to X-ray your hand. Berconi's in the hospital there and I'd like to look in on him. How about shitface there? Ah, hell, he can get his own X-ray. Come on Russky, we'd better clear off before he come all the way round. I've seen him fight and he can go. You can rest on your sucker-shot KO."

"Well," Jesse put in, "I wouldn't worry too much about the sucker-punch part of it. Didn't Mariano say 'Sportsmanship is for losers'?"

Randy was conscious and listening to most of this, but had enough on his plate without compounding his other troubles with a fight.

He pretended to be dazed to forgo escalating the mess. Anyway, he had three quiet days to sleep and practice ahead. It all simply confirmed his belief that the world was a shitty place and one had to take what he could any way he could.

Raining again—finally, Jesse thought. He was glad it had held back until evening. "Tomorrow may not be weirder but it certainly will be busier." He was to be in three finals. Sitting there in the bar sipping on a lager, he felt relaxed, but a tightness would irritatingly come and go. Karen had the car and was to pick him up. He thought for a moment of Shep, at that moment winging across the ocean, following the sunset. If everything worked out all right in the States, he figured Shep would be back Tuesday morning. That thought made Jesse feel better. Because with Shep back and another obstacle not cheaply removed from his way to the championship, it would be a worthy achievement. He didn't want any flukes sitting beside him in the record book.

He looked at Joe, the bartender. "Here's to it, mate. Crazy life, isn't it?"

Shep stared vacantly and red-eyed down on the banks of clouds; they looked like a vast range of snow-capped mountains. The clouds boded ill for Wimbledon, he thought perfunctorily, not knowing that the weather usually came from the opposite direction. Wimbledon hadn't lost its importance to him. But one reason for its significance had altered. Of the five times he'd played there, his father had seen only one. Money was tight, and Shep had always been an early-round loser. This Wimbledon, Shep felt sure, was his, and he and his dad had made elaborate plans—things that had for one reason or other, been postponed through the years. Now . . .

His eyes blurred again with tears and he didn't hear the question at first. "Sorry?"

The attractive blonde, in her late thirties or early forties, had noticed the tears. "Are you going to New York? Is that where you live?" Her Slavic accent reminded him of Eduarde Vector.

"Eh—no, m'am," he said hoarsely. "I'm continuing on to California."

"Not a pleasant journey?"

"No," he answered, not going into detail. She didn't pursue it.

When, fifteen minutes later, she rang for an iced vodka, Shep was more controlled and had his third scotch. They were becoming relaxed by the drinks, the hum of the plane, and the surreal "tube in space" effect of the situation, and small intimacies began to bubble to the top in their conversation.

She was the wife of the ambassador from one of the Communist countries en route to Washington. As a hard-core ideologue, she had that strained, severe look of intensity that zealots permit to freeze on their faces. Only when she laughed (more and more frequently after the second vodka) did the serious, brittle façade crinkle and fall away, revealing the glowing ashes of a beautiful face. Her short, blond hair brought to mind for Shep the Garbo film about a female commissar, *Ninotchka*. Shep addressed her in fun and a sort of ambivalent friendship as his "little commissar." It annoyed her.

"I *am* a commissar," she declared, slapping on the mask of tragedy again. Shep went back to calling her Ninochtka, after childishly referring to her one last time as "little commissar" to show that he wasn't intimidated.

The plane droned on through the brilliant sun of the upper spheres. Shep drew the shades and turned to her, saying, "Are you glad to be going to the United States?"

"Naturally I prefer my own country, but the United States can be amusing if I avert my eyes to all the social evils, the injustices, and don't get mugged."

"You must admit that, for all its foibles and errors, the people, in the main, run the government. In your country, the government runs the people."

"Of course. I agree. That is how it should be. Not ideally, but by the imperative nature of things."

Rather than jump all over that statement, Shep sidetracked the conversation. "I'm sure that you and I think differently about traveling, about this flight even. You are going from one country and set of social laws to another. For me, the world—I guess, because of tennis, it came from moving so easily and so often in a light frame of mind—the world is like a big house. I merely go from room to room in it. Of course, some of the rooms are locked up and when I do have the key, it's like visiting friends in quarantine. I mean—I can come back out again if I don't get too close to them. But all the rest of the rooms are easy. They have different sizes and flavors, naturally. All bedrooms differ and the kitchen and music room aren't anything alike, but they're all in the same house."

"That is charming and grossly simplistic. The people in some of those rooms are denied what some in the rest of the house have. There are also servants' rooms."

"Let's not belabor that one. You are on very shaky ground as far as I'm concerned, pointing the finger when it comes to equality and personal freedoms."

More silence. During some heavy turbulence, she clutched his hand and they were very close for several minutes. A moment or two after the plane smoothed out, she snatched it away again, slightly embarrassed. And just in time. One of her countrymen made his way up from the tourist section to see that she was all right.

"Who was that?" Shep asked, in a tone of one sharing confidences. Her voice was cold and busineslike when she answered, "One of the secretaries in my entourage. There are seven of them returning to their work."

"And they're flying second class and you're up here. I see." Shep couldn't, and didn't try to, suppress a light laugh. The laugh iced the conversation for several hours. They both slept until awakened for dinner. Shep glanced down at the shapely calf that merged softly with her boot, and felt guilty about a twinge of horniness that began to burn his sensibilities at he edges. It was years later that he found out that a sudden thrust of passion is a common occurrence when there has been a death or one is imminent among close family. It is a psycho-biological manifestation of a wish for survival of species; to procreate, to fill the loss, and to prove one's own existence All Shep knew at the moment was that he wanted to fuck Ninotchka in the lavatory. Not in New York, not in Washington, or some rendezvous somewhere else. Right now and in the toilet, sitting on the closed seat with her straddling him. Some of the heat or parasensory waves must have been communicated to her. She glanced hotly and fearfully at him. He took her hand again. Gradually, he moved it to his thigh. It surprised him a little that she took the initiative and slid her hand under his newspapers and grabbed him firmly. She held a slightly shaking *Time* magazine in the other hand, pretending to be engrossed in it.

Shep picked a blanket up off the floor and spread it across their knees. Airplanes were becoming very exciting places for him. He unzipped his trousers with difficulty and Ninotchka almost removed her hand; it hesitated, then took the plunge inside. He closed his eyes, and for a few minutes didn't give a shit about Wimbledon or Communism or Life or Death. Shep believed that the only time a man was truly, completely sane was one minute after an orgasm. There was a fine line between sanity and insanity and Ninotchka had just erased it with the last throbbing jerk. He sat there, eyes closed in his quivering sanity. Luck was good to them again. At that moment a secretary appeared from behind and a stew was coming down the aisle the other way with liqueurs. Shep's closed eyes and heavy breathing seemed sleep. She said something in her language to the man and he left. Very deftly, she wiped her hand on the inside of the blanket and accepted two Drambuies from the hostess.

When the hostess had gone, Shep bundled up the blanket and shoved it under the seat of a short, fat man with glasses and a fez in front of them. She handed him a drink, clinked glasses with him, and said, "It's my fortieth birthday today. But inside this middle-aged body, I am still nineteen."

"Happy birthday, my little commissar." He kissed her ear. She pulled back slightly, looking over her shoulder, then smiled at him with her eyes over the rim of the glass. Shep never saw her again.

The wake was stifling—lugubrious and false. There were relatives Shep barely recalled (even with the help of a resident spotter), ex-students and faculty, and LTA friends all working hard to generate sadness in themselves. Oddly enough, the LTA people, from whom Michael Shepard had long been estranged, were sincerely showing a sense of loss. Even outside their fold, he had been a force for good in the tennis world, and they appreciated his efforts while disagreeing with his methods.

Shep's mother was impassively aloof from everyone In a sense,

Mike Shepard had died for her the day he moved out of their home and into the shack up the coast with the twenty-year-old student. She knew in a textbook fashion that it was an aberrant reaction to the futility he felt—the sense of unfulfillment in the sphere he moved in, and the panic of onrushing age that short-circuited his logic. She knew all that, but never mentally digested it enough to understand it. They had been divorced three years that week.

Shep stared at the mercifully closed coffin. It was Saturday evening. His father and he would have been celebrating his win or discussing why he lost at Queen's, making plans for the Wimbledon weeks in the flat that Shep had rented for him in Chelsea; if his heart hadn't exploded. How absurd it all was. Shep stared at the coffin until he could see right through it into the past. Not a dim past, but a vivid, bright one. . . .

An eighteen-year-old all-American tennis player with a full head of hair, Francis Shepard took a step toward the net and nodded at the announcement of his name, Sammy Kerwin did the same from the other end. The Junior Championship was being played in the beautiful, specially adapted Rose Bowl for this now-major sports event, and pennants were snapping, vendors were vending, and not a few hearts were beating with an extra flutter from the adrenaline.

Shep smiled over at his dad, who was showing the strain far more than he. In the warm-up, Shep hit the first six practice serves into the net. Mike Shepard tipped his hat over his eyes and scrunched down in the seat.

The first games were played slightly nervously; they both were feeling out the right tack to take, looking for weaknesses. The day was the hottest and the brightest Mike Shepard could remember. The paper that evening confirmed it to be a heat record for Pasadena. The simulated cement surface radiated up to 114 degrees at the hottest time of the match. In spite of the fast surface, the rallies were long and hard-fought, some points going to fourteen and fifteen strokes. At 4-all, Kerwin was down with a cramp. The pain distorted his normally pleasant, Irish-Jewish face. He kicked his leg out straight and bent his toes upward while massaging his leg. Shep moved into the shade. Two linesmen and a doctor from one of the boxes helped him with the leg—and five minutes went by.

Mike Shepard called out to the umpire, "How much time are you going to give him? Be fair! Play is supposed to be continuous."

The umpire turned to him and, recognizing him at once, smiled uncertainly. "He's okay now. He'll be up in a minute." Six minutes later, Sammy finally was up on his feet and limping around the baseline, jiggling his calf and thigh with short shakes. The long break and the distraction of a hobbling opponent shattered Shep, and he lost the first set, 6-4. He hadn't yet the experience or instinct to jump on a God-given advantage.

Battling in the heat, back and forth, Shep won the second set, but just: 7-5. In the third set, a fully recovered Kerwin took control of the match and ran the score to 5-1 in very quick order. Suddenly, Shep sat down on the court.

"I've got a cramp," he declared.

"All right," the umpire said, "shake it loose and continue."

Shep replied, holding his leg, "No, this cramp is too bad. I can't play. You'll have to disqualify me." He wouldn't get up and waved off the doctor who made an attempt at helping him.

Now the officials were in a wonderful quandary. They were in trouble and they knew it and, for a few minutes, they were petrified with inaction. Finally, the referee made the announcment. "Ladies and gentlemen. Having consulted the rule book and conferred with my honored colleagues, we have decided that, however belatedly—in order to follow the strict letter of the law in application of the tennis rules, we must award this match to—Francis Shepard!"

The roar, the outcry, the rumble from the crowd was disbelief, the enchantment of surprise and, for some, partisan anger. Most just didn't understand what was going on or why. After the annoucement, Shep stood up and walked in gingerly fashion to the chair. Brian Kerwin, enormous, square-jawed, and enraged, exploded from the crowd onto the court, swearing and wildly waving his arms. "You fucking crooks! Oh, you motherfuckers!"

Mike Shepard was shaking his son's hand. "Francis—if I had a law firm, I'd be delighted to have you in it." Suddenly, Mike was lifted up into the air.

"Don't shake that little piss-ant's hand! He didn't win anything! He stole it!"

"That piss-ant's my son. Put me down, you ape!"

"Your son, hey?" Kerwin squeezed the lapels tighter around Mike Shepard's throat. Kerwin never saw the punch loop around his arm and pound him in the ear. He dropped Shepard, momentarily stunned, but reached out to grab and punch him. The police intervened, frog-walked Kerwin out of the stadium and into a police car. They had to lock him up for several hours before he was cool enough to release.

Through all of this, Sammy just watched with a forlorn expression of personal heartbreak. He turned to Shep and, taking him by the shoulder out to the middle of the court, shook hands with him. The stadium vibrated with applause—everyone on their feet, appreciating the incredible gesture of sportsmanship. There was no need for words between the two. Sammy's father was infamous. He was the archetypal tennis parent whose scenes of unfathomable hysteria, outrageous behavior, traveling to every known tournament, made him the scourge of junior tennis. The adjectives rolled off the tongue at the mention of his name, and anecdotes, some untrue, were reported from all corners of the tennis world. It was a tribute to the boy's mother that his sanity wasn't totally frazzled by the crushing insistence on performance.

To make it, if possible, even more embarrassing, Shep and Sammy were partners the next day in the National Doubles Final. Shell-shocked as he was by a lifetime of personal horrors, Sam was still capable of embarrassment. His chin was on his chest as they walked off into the tunnel to the locker rooms. Shep wasn't all that proud of his "cramp" tactic either, but they both knew, understood, that Sammy should have been disqualified in the first place.

Sunday morning unrolled a rare day. The smog was low, the sun not fierce but embroidered into the slightly impressionistic haze of blue. Tennis fans had been lined up for general admission tickets for hours, and at one-thirty, Shep and his dad swung through the gates at a vigorous pace, Mike Shepard needing two steps to Shep's one. They had been warming up at the Anandale Country Club, and Shep felt in first-class shape, especially mentally. He was eighteen and the world was a toy. The match was due to begin at two-thirty. At two Sammy hadn't yet arrived. At two-fifteen a worried Shep gave a relieved whoop when finally Sammy hangdogged into the locker.

"Sammy, where've you been? Get ready."

"I can't. My dad hid my clothes and broke all my racquets. He locked me in my room, but I climbed out on the roof and down a tree."

"Dad!" Shep shouted. Mike Shepard appeared from the lounge with a mug of beer. "Sammy, my boy. There you are. Don't tell me —he won't let you play? He took your stuff, right? Oh, Christ, I am right. Okay—shoes—Shep?"

"I'm size twelve, he's eleven."

"Give him the extra pair. Wear two pairs of socks, Sam. Shirts we have. Can you use one of Shep's racquets? Never mind; you just have to. You play with Wilson anyway, don't you?"

"Yeah, but board tight. Shep has his at fifty-four pounds."

"There's a lucky break," Shep exclaimed. "I have one that's too tight. They just strung it for me and goofed it up. I was going to rub it with a candle to loosen it."

Mike Shepard was warming to the crisis. "I know where I can get some shorts. I saw a guy your size outside walking around pretending to be a player. For ten bucks and the glory of being a part of the match . . . well, let's see. I'll be right back."

Sammy got into the emergency mood, too, and the scene became a hectic, interesting team effort. He stuffed toilet paper in the toes of the Adidas shoes and put on Shep's spare pair of socks over his street socks.

Mike came cantering in waving the shorts like a scalp. "Here they are! He's going to watch from the press box. I arranged it with Bud Walker. God! a jock—you need a jock!"

"He can wear my jockey shorts. They probably make more sense anyway."

"The show's going on," Mike said, beaming.

"So far. Keep your fingers crossed, folks," Sammy said, his enthusiasm flagging. They didn't have long enough to wonder what Sammy meant. The boys walked out on court through the middle tunnel—and Brian Kerwin was heading right for them from the opposite direction. He had begun to bellow from the other portal. "You're not playing! You're not playing with those cheaters!"

"Oh yes I am! Leave me alone! If you hit me again, I'll flatten you. I don't care what I promised Mom." Sammy's sad-eyed, pretty, prematurely aged Jewish mother had spent her entire married life refereeing the two, trying to keep the boy alive in the early years and, now, attempting to keep Sammy, who had become the stronger, from wreaking retribution.

Kerwin made it only halfway across when the police, pre-alerted by Shepard, intercepted. He had been drinking and put on a fine exhibition for the crowd, who were reveling in this unexpected warm-up match. When, finally, he had been removed, showering the proceedings with barracks invective over his shoulder, the gallery settled down to tennis. Sammy Kerwin and Francis Shepard, one hour and forty minutes later, were the junior doubles champions of the United States.

Back in the locker room, Sammy returned the gear (the boy in the press box, forgotten, remained there until dark), borrowed fifty dollars from Mike Shepard, and left to join the Marine Corps. One of the best, most promising players in the country was thus lost for three years.

The dredged-up daydream dissolved, ripping with it the sweet, silken membrane of escape. Shep was hurled back into the reality of the moment with a wrenching shock. The coffin was again a coffin. His father was dead for him at this moment, in realization as well as fact. He glanced around at all the morbid faces hanging over dark clothes and stood up abruptly, saying in low tones:

"Everybody . . . friends and relatives . . . thank you for coming . . for your prayers . . . your good thoughts. I know that my dad's at peace. Now my mother and I would like to get some too, so I'm going to close the wake at this sensible hour. Anyone who comes later will have the satisfaction of knowing they will get Karma points for their good intentions. Good night."

"Francis! What are you doing?" his mother whispered.

"Come on, Mom, we're going out for a drink. I have a surprise for you."

The group broke up with a nice parley of dismay and relief. Shep shook hands with everyone in the large room as they left and turned out the lights. He knew there was no one in there, but found himself forced to say, "Good night forever, my dad."

They were a quiet pair in the small booth of the Sandpiper. Without his father there to give it dimension, Laguna seemed, for the first time, very small.

"You have packing to do tonight, Mother. I have reservations for both of us tomorrow right after the burial service. You're coming back to England with me . . . eh-eh . . . don't argue. I'm going to win this Wimbledon for Mike Shepard. For you too, Mom. And to tell you the truth, I wouldn't mind winning for myself just a little."

"But really, Francis—why not wait a few days? I haven't any clothes to wear to London."

"We can be in England by eight on Monday morning. Then I can sleep off some of the jet lag and still practice in the evening. It stays light till nine at least. As far as clothes, I must be, oh, probably four Mother's Days behind. We can shop in London. No weedy stuff—young colors. We have lots of living to do, Mom, old girl."

Her eyes acquired a new sparkle. Or was it something returned after a long absence. Sylvia Braden Shepard found a new infusion of positive energy on an evening she had least expected one. She drank a neat whiskey and followed it with a chaser of soda water.

"Mom, you're supposed to mix those."

"I know, Francis, I know. Just this once. I want a quick high. Then I'll sleep like the dead." In her happiness, she didn't even notice the expression.

"Probably you'll find a suave, lovable Englishman. You're a young-ish lady. We're going to live for three now. Tell me. Do you forgive him?"

"I suppose I do. I guess everything is forgiven in death. Yes. Of course I do. But you are most certainly right. It's my turn to break out of my rut. How did that Thoreau passage go? He said he finally left the pond, the woods, because it seemed that he had several more lives to live. What was it? 'It is remarkable how easily, and insensibly, we fall into a particular route . . .' Then something, something, and . . . 'how deep the ruts of tradition and conformity . . .' Thank you, Francis. I like you. You're a good person. Make a toast and let's get some sleep. I know I'm going to cry tomorrow. The grave's the thing that gets me."

"In that case, here's my toast. They say that the world is a huge airport where souls are perpetually landing and taking off. Here's to Dad's trip. Cheers and Godspeed. They are only burying his suit tomorrow."

The following evening, Shep was crossing the Atlantic for the third time in about a week. It was Sylvia Shepard's first time. A note was passed to him by a hostess; her familiar form whisked through the curtain into first class. It read: "Get rid of the old lady and I'll meet you at the Dorchester?" Shep thought: *Inshallah.* Life goes on.

.

24

Names of cities, Hiltons, Ramada Inns, of stadiums, halls, matches, all blended together over the years. The girls were Picasso cubistic faces, floating transparencies over one another in a whirlpool of arms and legs. For Shep, some phrases stuck together like memory pegs, sharp and timeless: Philadelphia, Fernbergers, two semifinals, Chicago, Danielle, DePaulis Alumni Hall, Little Rock, Marylou, Houston, a win at the River Oaks and a billboard by the airport in huge letters that said HUGHES TOOL WORKS, underneath which, in red paint, someone had arrogantly painted "So does mine." Dallas, three qualifications, and one win in the WCT. Finals—his respect for Lamar Hunt, the humor of Mike Davies, Las Vegas, Caesar's Palace, hospitality, two wins, and the warmth of Alan King, also his first professional contract signing, no . . . he did that in L.A.

Even the important things were beginning to fade in the jumble of similarity.

His mother was asleep in the seat beside him. She had just relieved herself of another Thoreau slice of wisdom. Maybe it applied to her, but not, so far as he could see, to him: "I have always been regretting that I was not as wise as the day I was born. The intellect is a cleaver; it discerns and drifts its way into the secret of things."

Shep couldn't quite grasp why she had told him that—at that special moment. He'd ask her later. Better to clear his mind completely. His first Wimbledon match was that afternoon. No postponement. He was sick with nerves and doubts. Jet lag—three days without practice and an opening round against Renato Bovino, the Italian number two. Shep remembered the big wink Berconi had given him. Their visit had been almost wordless at the hospital, but as Shep was about to leave, Berconi turned his head toward him. The wink was unmistakable. Now, Shep almost envied his detachedness from it all. "Goddamnit! This should be my Wimbledon." He looked at the crinkled cable again.

SORRY NO POSTPONEMENT POSSIBLE FIRST ROUND MONDAY BOVINO

THE COMMITTEE

Just maybe he was in luck. Outside the window of the plane, England, as Wick would say, was raining like a tall cow pissing on a flat rock. It was only seven in the morning and it could stop, but it cheered him a little. He woke his mother.

Once they were installed in the hotel, and Shep had been able to sleep four hours, he had a courtesy car pick him up and whisk him out to Queen's. The LTA and Wimbledon referee had reserved court one for him for the entire day, a prodigious exception, considering everyone else was allotted a half hour. He soon found out why.

The office was tuna chunk solid with people asking, demanding, and explaining things. He squeezed over to the bulletin board.

The list of seeds:
1—Jesse Fraser
3—Jaguar Gray
5—Jean-Claude Dumée
7—Al Wick
9—Freddie Moore
2—Francis Shepard
4—Randy Mariano
6—Billy Sherman
8—Paul Owibakum
10—Buster Matthews

218

11—Luis de Leon   12—Pancho O'Brien
13—Dan Sullivan   14—Carl Zawitsky
15—Mikhail Pakachev   16—François Joffre

Number two seed. His confidence swelled again. "I can do it if I can only get through today. I can do it." He dashed up the steps to the changing room and dressed quickly. Billy Sherman was tightening a new grip around his racquet handle on the bench beside him.

"Sorry about your father, Shep."

"Thank you, buddy, thanks." He quickly changed the subject. "Who did you draw in the first round?"

"I play your fat friend, Sammy Kerwin. He did well in the qualifications. I shouldn't have any trouble with him, though. Five sets is quite a few when you're overweight."

"Don't take him for granted. He was a hell of a player and he doesn't look so out of shape to me. Who won Queen's?"

"Jesse and Jag were a set all and 4-4 when the rain bucketed down. They had to wait three hours for it to stop, and when they came out Jag was stiff and lost his serve. He almost broke back but Jesse walked off with it after all. Good match."

"And the girls?"

"Laurie beat Missy. Very close—8-7, 7-5."

"Billy, could you possibly hit with me for a half an hour or so? I haven't touched a racquet in days—you know, since Friday."

"I really am very sorry. I really am, actually, but I've just come off, I have a two-o'clock match, and haven't had lunch yet. I know bloody well you're nervous about your match, but I don't want to lose mine either, you see. Sorry about that." His English presentation made the intensity of his concern sound legitimate—and it probably was.

"That's okay. There must be someone around. I just want to get some feel."

"Oh, yes, sure, Shep. You'll find somebody. All the best."

Shep asked several other people. All were going to lunch. Was he going to be reduced to hitting on the wall? He went downstairs, the foam of panic forming in his mouth. Two more polite, apologetic rejections. "What the hell is this, *High Noon?*" Finally, at one o'clock, he found a young junior who was so astonished, so awestricken to be invited by Shep that the poor boy couldn't hit a ball right. Jean-Claude, happening by on his way to lunch, saw Shep's problem and replaced the terrified youngster. Shep's warmth of appreciation glowed in his smile. "Thank you. *Merci bien,* Jean-Claude."

"Pas de quoi, mon vieux. I can't hit very long but, *mon Dieu,* it is impossible we have the second seed walking cold to center court. I have easy draw, *moi."*

It was less than adequate, but better than nothing. At two-thirty Jean-Claude and Shep shared a Rothman's courtesy cab out to Southfields, home of the All-England Club, better known as Wimbledon. The heavy traffic flow across Putney Bridge only heightened the tension and tightened the knot in Shep's stomach. His head ached from an upset sleep pattern and metabolism upheaval. Jean-Claude prattled away, apparently oblivious to Shep's nerves but actually very aware—trying to get Shep's mind off the impending torture. Its effect was the opposite. Shep's nervousness increased in direct proportion to Jean-Claude's levity. On the other hand, Shep didn't want to say anything about it, considering the great kindness Jean-Claude showed with his help. He had to forgo lunch and they were both eating turkey sandwiches out of brown bags and drinking canned Coke.

The car was ushered through the gates with a flourish to the applause of the ticket line and those milling around inside. They drove slowly, cautiously through the thronging crowd of opening day—past the giant green scoreboard, beneath the players' enclosure balcony and bar, through gaily colored hats, strawboaters of schoolgirls, bright dazzling dresses and elegant suits, faces trying to peer in, and waving in recognition.

The flowers were copious, enormous, and very cheerful. Shep relaxed a little. The car slowed to a halt in front of the players' dressing rooms. As Shep and Jean-Claude stepped out, a terrific roar from center court augmented the handclaps they were receiving from the autograph hunters around the door. The All-England pennant fluttered with quiet dignity above their heads, outside the sacrosanct committee rooms. It was a strawberries-and-cream, picnic-basket day, but the canny natives through whom Shep made his way carried umbrellas and raincoats over their arms. This was London, after all, and one never knew.

It was standing-room-only on the outside courts, and the electricity of hopes and aspirations being laid on the line in pure head-to-head ability, combined with the partisan energies of the crowd to set up great waves of force and excitement all over the sunny, green-checkered landscape. The aroma of overcooked Wimpys lent their little stamp of validity to the composition. The queues for the hamburgers were long and motley. The world's largest tennis garden party was set, and the three hundred thousand fans who would pass through the gates were, although their voices were modulated, as nervous and excited as the contestants. The vortex that was Wimbledon was pulling the tennis world into itself for two weeks.

Some of the players were watching the big color television set in the locker room. Jesse on center court was romping a seasoned Spaniard and looked like a straight-set victor. Shep had no time to lose changing.

"I am so glad we aren't the first matches on," Jean-Claude said, with one foot up on the bench, tying his laces. "Do you know that the umpires they have their cocktail party at twelve-midi opening day? Even if they go, how you say, lightly on gin and toniques, for sure there must be some very interesting calls for the first hours or so. Not so many years ago I remember one lady madame umpire she fell asleep on center court. The chair umpire he had to wake her up to ask a decision."

Shep nodded, only half listening. He was thinking about the surface. The ground under Wimbledon's grass is hard, like cement. The Forest Hills soil is soft, Queen's Club is spongy, and outside of the fast wood practice that he'd had, Shep knew that the five-minute warm-up, knock-up, as they called it, wouldn't prepare him for the low skids.

Bovino was dressed already and watching Jesse's match. He knew that, ordinarily, his chances wouldn't have been much, but the odd conditions that afflicted Shep gave him a better shot at it. On the other hand, he wasn't happy with the grass either—considered it very passé.

A service ace, a roar from the crowd, and the first match was over. Renato Bovino blessed himself and joined Shep standing by the door, waiting as the eleven new linesmen trooped single file out into the arena, their blazers bearing badges of authority heavy with importance for the Wimbledon weeks. Some of them were a bit beyond middle age and the term "Dad's Army" was coined for them after a popular TV comedy program about the British Home Guard. Once they were in place, the two players walked out side by side, turned, and gave short bows toward the royal box, then continued to the umpire's chair and unloaded their racquets and gear. Bovino won the spin. Shep looked around. He'd played here often before, but it still seemed enormous to him. He found it hard to believe that it was a regulation court. A light plane seemed capable of landing there. His first match, first big test toward winning the big one was on.

The players' cafeteria was swarming with guests, past greats—Talbert, Hoad, Budge, Seixas, Santana, Ashe, who had just retired last year—and a sprinkle of players. Wick walked over to a group of players, careful not to tip his tray and spill anything on his sharply pressed white linen suit.

"You look cool, Wicko," Buster Mathews greeted him.

"You don't look so hot yourself," Wick grunted, in obvious bad humor. "This food would gag a maggot."

Paul Bell brightened at Wick's salty mood. He thought that Al was a very funny guy. "Ho, Wicko! How's it goin'?"

"How's it going yourself? How's the ankle?"

"Be fit about Ireland," he answered, referring to the Irish tournament after Wimbledon.

"Sorry, mate. That's very hard luck indeed. You made the right decision not to play on it, though. I see Mariano's still consorting with those seedy gangster types. He's going to end up in a concrete canoe if he's not careful. You can't fool with those babies. If he pulls some of his crap on them, they won't even be able to gather him up with a rake."

Jesse came in and joined Karen, who kissed him. The kiss was brief but meaningful. One down, it implied. "Chroist, Wicko. By the looks of the scoreboard, you didn't have any easy time of it. Four sets—two tie breaks. You didn't go on the piss the night before Wimbledon, did you?"

"Of course not. That bloody Gohon just played great tennis, that's what. Don't underestimate those Africans. Just give them a couple of years. They'll be wiping us up."

Jesse said hello to Paul and Buster. Everyone mumbled their

congratulations. It was pretty much a foregone conclusion, his win. The courtesy was automatic.

Paul had limped to the bar to fetch three lagers and returned with them and the news that Shep had just lost the first set.

"I'm not surprised," Wick remarked. "He's not fit. It's a wonder he's here at all."

Billy Sherman wandered in, still in a daze over his loss to Kerwin. He pulled up a chair and sat down among the murmured litany of condolences. He summed it all up by admitting, "What can I say? He just outplayed me. Before the match, I was calling him the Count of Monte Crisco, you know, fat in the can, but he played like a buzz saw—never let me in the match."

"Never mind, Billy. Cheer up."

Sherman didn't look cheered up. His smile was constructed with great effort. "I didn't fancy my chances of winning the whole thing," he said, "but I just wish that I could have justified my seeding—made it to the last sixteen at least."

"I went out the first round the year I was defending champion," Jesse chimed in, "so if you think you feel bad, try to imagine that!" Billy couldn't and didn't attempt to. He was tired of maintaining a brave face. With a "well done" to the others, he went into the players' bar to wet down his disappointment.

Wick told jokes and otherwise entertained the lunch group. They watched the final scores being posted on the scoreboard and caught bits of the matches being staged on courts two and three on either side of it. Karen left the table on a tea run as Laurie joined them. Jesse made a great catch of a salt shaker before it hit the floor. The tables around them gave him a round of applause. He stood up and bowed. Laurie was uncertain whether she should sit down or not. "What's the big exodus over here? I just came over to tell you that Shep won the second set, and people are running off in all directions."

They feigned an interest to match her enthusiasm. It was clear to all of them that she had a more than academic interest in the outcome. Laurie was in love. As always, it was a little uncomfortable for them, this exhibition of a soul lost; for everyone except Karen, the romantic, who had returned. Karen had never allowed herself the luxury of much romance in her early life, and indulged herself whenever she saw others enmeshed in the bittersweet morass. She watched Laurie's eyes as she described the set. Like neon signs, they blew her cover. Karen broke the embarrassed silence saying, "Why don't we go out to the players' section and watch some of it? Jesse is going to practice some more, if you can believe that." She took Laurie's arm as they walked to the stairs. "He wants to win this one with all his heart."

"I guess you know that I want it for Shep, but it shouldn't interfere with our being friends, should it? We don't have much influence over the outcome one way or the other."

"That's the way I feel, too." Karen called back to Jesse, "See you about seven at the flat, Jess. Don't fall down or pull anything."

Jesse threw her a kiss and said, "That's right, love, think positively. See you."

Except for Sherman, all the other seeds made it through. Shep won in four sets, 6-1 in the last. It was a very strong, impressive finish. He and Laurie drove back to London together, stopping for ice cream on the way. Shep didn't discuss his match much, but his occasional deep sighs were broadcasts of his relief. A day off, some grooving workouts, and he would go into the next round against the Czech number one with more confidence. Taking life day by day, as he had conditioned himself to, made the living of it much more manageable. The twenty minutes of meditation in the mornings and evenings gave him a gestalten unity to things but couldn't completely erase his anxiety. He wanted this very badly. Wimbledon. The great one. For the moment, he was savoring the sweet evening and the mellow mood of Laurie. For her part, Laurie was wrestling with the dilemma of whether to tell him her feelings and thus risk rejection and the possibility of intruding on the total concentration he required to win. She understood how much he desired the championship—this one above all others; she wanted it for herself with every cell in her being. The fleeting thought, the possibility that they might both pull it off was rudely banished; the jinx raised its ugly, serpentine head.

Shep didn't have to be told. The small glances, touches, the little extra kindnesses she performed said it for her. For a time, Shep had shared Jaguar's normal state of confusion. Eventually, he decided to roll along with it, and either take it broadside or sidestep it when the proper time arrived. But he couldn't help shaking his head, wondering why, of all times, it had to happen to them now. For the time being he opted to simply gloss it.

"Did you happen to meet Sammy Kerwin's father up in the tearoom?"

"Not actually meet him," Laurie answered, "but I saw him. He didn't look like the monster in all the childhood stories I heard about him."

"That's just it. He's a new man. Quiet, thoughtful . . . and sober. I suppose that some of his fanatic passion ebbed during the years when Sammy was a tennis dropout. He's in AA now. The blood stands a better chance of reaching the brain nowadays. Jesus, he is really a nice guy—hard for me to believe it's the same person. Sammy is so proud of him."

"He should be proud of Sam, too, after today. What an upset."

"Now, you dig, I don't consider it an upset. I know the guy and he has more natural talent than eighty per cent of the field. Watch him."

During dinner, in the small, intimate, seven-table restaurant, their conversation narrowed into a strategy conference. Laurie's opening match was to be against a tall, willowy blonde, a German perennial who was a giant-killer, solid off the ground with a strong, reliable serve. Shep was good for Laurie, infusing her with strength and positivism, not to mention some damn helpful tips. "I've noticed that you don't get a high percentage of your first serves into the ad court. Don't ever forget that when you're serving to the backhand in that court, the net's much higher. Throw it—make your throw a little higher and watch it a tad longer. If she tries to drop-shot you, and she will, plenty, don't try to drop-shot back. Don't dick around with it. Go deep for a line and drop back a couple of paces. She'll have to lob from there."

Laurie was so brimming over with excitement and the fullness of the moment that she had no emotional brakes left. "I love you, Shep. I know. I know I shouldn't say that . . . it must sound completely nuts to you. No, please . . . don't say anything. Just . . . don't pay any attention to it. Don't let it upset you any. I can handle it, I'm not silly . . . not the gushy, clinging type. Whatever feelings you have . . . if you have any at all for me . . . examine them later

. . . some other time. Now, let's relax, if we can after all this, and enjoy the rest of the evening." Laurie was flushed crimson. The color heightened her lovely face and Shep felt required to say something.

"All pretty sudden, isn't it? I mean . . . Jesus . . . I . . . I don't know how to react. Oh, I could do a bullshit number in here. But that's not fair. I've always lived and thought . . . worked on the premise that no one really cared about me. That they were only indulging some emotion toward me for whatever feedback they could get. Even my dad was getting his jollies vicariously through me."

"I care very much what happens to you, Shep. It isn't sudden. It was the old 'admired from afar' thing with me. But sometimes the . . . when you were near me, spoke to me . . . the sound of your voice . . . like now . . . the sound of your voice obliterates the message and my mind sings and forgets to listen." The lyrical quality of her quiet, tuneless love song and the obvious sincerity in her whole manner embarrassed Shep. He felt unworthy. The thought of love between Laurie and himself had simply never existed. He knew, of course, that usually one is loved and the other is active, but he wasn't quite sure he was capable or ready for the responsibility of it all. This girl was serious. She could tell by his eyes the quandary she had hurled him into.

"Hey, Shep, please darling, let's . . . don't get yourself strung out trying to figure it. I don't want to upset you. Especially now. Sometimes I atomize my anxieties and spray them all around me. I shouldn't have said anything, goddamnit!"

Shep didn't want her to be shaken up either before an opening match. He took both of her hands in his and, leaning across the table, kissed her, full, long, and intensely. For the moment, there was only the sound of their breathing, captured and magnified in the trembling silence. At the end of it, they sat nose to nose, thinking far into each other's eyes. "Let's go home," he whispered.

25

Angie stepped down from the train, making two loud clonks with her mod clogs. The chauffeur of Señor Jorge Gainza-Keppler took the two bags that the conductor handed down and toddled after her toward the waiting Bentley. Since her visits had become more frequent, protocol had given way to relaxed familiarity, and neither Jorge nor Consuelo felt it necessary to meet her train every time. Angie was conducting a well-thought-out campaign to make Consuelo. She was hot for her, but she was keeping cool; constraining herself with steel discipline and, at the same time, playing the charade of conversion. She was pretending to go straight even to the point of sleeping with Jorge. Consuelo was only her tennis partner and confidante in the farce they were acting out. Whether consciously or not, both Consuelo and Jorge were aware of the confection; it was a game they all entered into in order to accentuate, to spice up the inevitable when it inevitably happened.

Angie was biding her time until Consuelo was ripe and heavy, ready to fall into her waiting hands. Too often she had plucked prematurely. Consuelo was far too lush a prize to mess it up this time.

The fifteen-minute drive from the station took them into the Oxfordshire farm country. Through tortuous, narrow roads, twisting their way under elms and English oaks that meshed branches above, canopying most of the way—between high banks like tobogann runs and thick hedges several feet thick, the Bentley hummed. Angie leaned back in the sumptuous comfort and closed her eyes, lulled and rocked by the smooth turnings.

The huge mansion was set back two hundred yards from the road, its stone masonry a total blend with the landscape. It had been built by a lieutenant of Warren Hastings with money siphoned out of the East India Company, and no expense had been niggled. It had strength and simplicity. Inside, it was lavish or garish, take your pick.

Jorge had rented the estate for the summer months as a headquarters while on business in England and as a redoubt for his cousin Consuelo, to give her a base of familiarity in her try at Wimbledon. The servants were Argentinean, and an old and loved Aunt was installed as duenna.

Jorge, Consuelo, and two neighbors were gathered by the pool. Jorge and the neighbor were drinking Sangria and Pimms Cup #2; Consuelo drank iced tea. Jorge, the complete Latin gentleman, sprang to his feet and kissed Angie on both cheeks.

"*Buenas tardes,* Angelina. How well you look. Will you have a cool drink?"

"Well, yes, thank you, Jorge. Hello everybody." Angie swung into the circled chaise longues and plopped down on the end of Consuelo's. "What are we having to drink there, Connie? Don't get loaded. We have work to do out there." She waved in the direction of the tennis court.

"Don't worry about me," Consuelo said, smiling. "We'll go and change in a few minutes."

Consuelo introduced the young English couple who were her

neighbors, three miles away, then led the way to show Angie her room.

"Jaguar's coming up for dinner tonight. Just the six of us. The Salem-Guests, Larry and Martha, will be joining us. Did you bring a dress?"

"Just the tailored suit I wore. I mean—I didn't know it was going to be so la-de-da."

"Well, don't worry about it. Wear whatever you like. You're in the same room as last time. You're comfortable there, no?"

"Oh sure. I'm comfortable there, yes. I love it. Why wouldn't I? It's divine. So—I'll see you in about ten minutes at the court?"

"No. Stop and get me in my room. I have something to show you."

"What is it?"

"Ten minutes then. *Ciao.*"

"Oh, all right, Captain Mysterious"

The room Angie had chosen was, not surprisingly, very masculine, with lithographs of racing horses and riding cups, ribbons, trophies, and leather trappings. When she had unpacked and zipped into her tennis dress, she stood in front of the mirror, posturing, trying to assume an attitude, neither feminine nor too butch. She gave it up as a bad show and went to Consuelo's room.

When she heard Angie's footsteps in the hall, Consuelo turned and with the back of her hand extended the diamond shining in the shaft of sunlight from the window. Angie's heart sank. It took a very deep breath to achieve any voice at all.

"My God! That's gorgeous. You mean he really—Jaguar actually —*proposed* to you? I don't believe it!" She ran to Consuelo and kissed her cheek, hugging her with a phony show of force and enthusiasm. "It's wonderful, Connie! Just super. I hope you'll be very happy."

"Perhaps I'm crazy," Consuelo said. "I know what a Don Juan Jag is, but I don't know . . . we've been having a very romantic affair for three months now. I imagine that it is not a secret any longer. He is terribly kind . . . gentle with me. He loves me. I believe that. And I . . . am quite mad for him. This is a trial engagement, I guess you would say. We'll announce, yes, but I have no fear, no hesitation. I'll change my mind if need be. But isn't it exciting?"

"Terribly!" Angie again forced a smile. "But come on. Let's go to work."

They walked the quarter mile through the deep, delicious, humus-smelling forest to the court. Angie made a decision. It was bold and made her heart race. For half an hour they hit corners, cross-courts, first on the backhand, then the forehand. For another half hour they took turns at net volleying. It was Sunday. Ladies' Day was Tuesday and the suspense started to grip them. They were both psyching themselves to be up for the first-round matches, experienced enough to know better than to take their weaker opponents for granted. They were both in the same bracket, creating the long-shot possibility of meeting in the semis, and, since they were doubles partners, their fortunes were thus intertwined to some extent.

Consuelo sailed back and forth across the court like a felucca, a delicate, graceful Nile boat. When she leapt to smash, she was more like a dove in flight than a power hitter, but the ball was blasted away nonetheless. Angie's desire gorged up in her throat, doubled and redoubled. Then Consuelo stopped and held her hand up to wait a moment. She walked toward the net and beckoned Angie to help her. It was obvious that she had something in her eye, a piece of grit or pollen or something. Angie leaned Consuelo back against the net and peered into the eye, holding it open with her left hand. She saw the little black gnat lodged just inside the lid near the corner. As she leaned forward, intent on getting it out, their breasts touched. They both stiffened for an instant, then relaxed. Angie pressed harder. Consuelo's breath quickened and Angie could see the pulse in her temple pick up in tempo, but she didn't slide away. For an instant, in fact, Angie could swear that there was an almost imperceptible return of pressure. As Angie inclined her face closer to scrutinize the eye, exaggeratedly close, she allowed her nose to brush Consuelo's. There was a spark, a quick shortness of breath in both of them. Impetuously, Angie kissed her, gently at first, then harder, darting her tongue rapid-fire into the startled, partially open mouth. Consuelo didn't pull away. Her heart pounding, she leaned wide-eyed back against the net. More confidently, Angie lifted the tennis shirt with her left hand and cupped Consuelo's breast, lifted the bra up to expose it, and slipped her right hand down into her shorts, beneath her panties and across the thick bush. Consuelo, in a trance, returned the kiss and raised her pelvis slightly to give Angie access to her now very wet pussy. Birds chirped and cicadas buzzed in the trees, all unheard. They felt as though their bodies were levitated in the beauty and purity of the moment. Seen from a camera in the sky, they were simply one other organism throbbing among the other life-forms of the forest.

Angie unzipped the shorts and dropped to her knees, pulling them down with her. She had already started kissing it before Consuelo stopped her, pulled her back up to her feet.

"Not . . . not here . . . gardeners!" she exhaled. With two quick motions she pulled her shorts up and rezipped them. Angie knew she had won. Silently, they gathered up the racquets and balls and walked back to the house, entering it through a side door. Without appearing to sneak, they crept, still in silence, to Consuelo's bedroom. Consuelo locked the door behind them.

One by one, the results were posted on the green scoreboard: Silverman, 6-2, 8-6; Teaford, 6-1, 6-3; Martin, 6-4, 6-4; Dubois, 8-6,6-1; Redfield, 6-2, 6-4, and so on. All the seeds came through. On Wednesday, news of Jaguar's and Consuelo's engagement sold papers to sports fans, and when they walked out on court two for their mixed doubles match, they received a standing ovation. It flustered and confused, thrilled and embarrassed Jag, who promptly contributed to the loss of the first set. A young girl in the stands fainted at the announcement. Reporters were trying to find out her name. Once the loving couple pulled themselves together, they were too much for the Japanese pair they were playing.

Anne-Marie Dubois was the only women's seed to be upset in her Thursday's match, and all of the men's ·seeds prevailed. Randy Mariano celebrated his victory by dropping another thousand pounds at the Brass Shoe. He went to a chip-hustler's apartment for a little tenderness. There were three men waiting in his hotel room when he returned the next morning.

Getting her period a week early—nerves, no doubt—released Laurie from the worries she had in that area. She was playing first-class tennis, but so was Missy, whom she was seeded to meet in the semis. They were both looking forward to it.

Saturday was wild and woolly. Jesse beat Joffre as expected, but Sammy Kerwin dumped Buster, and in a heartbreaker, Luis de Leon upset Wick. Al wasn't playing anywhere near form, but the bad breaks really closed the gap of class between them. With "chingados" echoing all over court one, de Leon soft-served, lobbed, and lucked his way to a win: 3-6, 8-7, 7-5, 8-7. Wick was devastated, flung into a deep depression that manifested itself as bluster and a false "easy come, easy go" flippancy. Privately, he considered not going to Jag's engagement party in Oxfordshire that night. Still, he rationalized that it would probably be the best tonic for him.

Freddie had a tough match with Owibakum but won. Shep had a straight-setter over Pakachev, or Billy Budd, as Shep now called him. Because Pak had not been able to think of the proper insult in English to reply to Mariano, he had socked him. To Shep, it was a classic response.

Mariano repeated his win over Danny Sullivan in a violent name-calling contest that so unnerved Sullivan that he turned into the "heavy," throwing his racquet and questioning everything. Mariano seemed tense and preoccupied, desperate at times, but it didn't cramp his intimidating style all that much. He got to Danny and the boy blew whatever chances he had.

Jean-Claude Dumée had an exciting squeaker against Pancho O'Brien, winning in five sets. It put him in his first quarter-final ever at Wimbledon. He had been a semifinalist two weeks before in France, losing eventually to Berconi, but that was at Roland Garros on the slow clay. Berconi had won there as well as in Rome, so Jean-Claude's showing was very respectable.

If Jaguar were as clear-minded and adept off the court as on it, he would be a combination of Bobby Fisher and Nureyev. He demolished poor Carl Zawitsky, 6-2, 6-2, 6-1, and had de Leon to play in the quarters, for him a relatively safe conclusion to be drawn.

Jesse was very empathetic, obviously upset by Wick's loss. He decided to avenge Wicko on Tuesday. On the other side of the ledger, the loss negated the distasteful chore of having to beat his own best friend. Things seemed to balance out.

Shep and Laurie decided to take a train to the party and get a lift back by someone instead of fighting the weekend traffic, the exodus to the country. They were very lucky to get the one taxi in Oxfordshire to take them out to the mansion. The driver was about to knock off for the night, business or not, quite typically. For a fiver he consented, eventually.

The party was casual and fairly subdued, considering what it might have been with the guests, in the main, tennis players. Jorge had hired a trio for background music and incidental dancing. It wasn't the bash some expected. The young were disappointed, the veterans relieved.

Shep observed to Laurie that the decor was overblown. "It reminds me of the Victor Emmanuel Monument in Rome. You know, the wedding cake? It's the old artistic principle that if one is good, six is better."

"It's like what I told my parents about their place here. Venetian Jewish."

"Yeah, that's good," Shep laughed. "That handles it."

The happy couple were randomly greeting the guests, receiving congratulations and good wishes. Consuelo was fighting back a potentially paralyzing doubt, but Jaguar was absorbing the moment, hanging right up there on the apogee of happiness; a pint in his hand, someone he loved on his arm, and the stars of goodwill in his eyes. He was in a daze of contentment, taking the general good cheer of the room for him on blind faith.

Wick was hovering by the piano, waving a beer with one hand, looping an arm around Billy's neck, and singing a song that had no similarity whatever to what the group was playing. After he had asked them to play "Take Your Love and Shove It," they had decided to ignore his presence. Wick turned to Billy and, indicating the new arrivals, Shep and Laurie, said, "And here comes superstar with Lady Silverman. She had her nose fixed, now her mouth doesn't work."

"That's rather uncharitable of you, old sod," Billy objected. "She's a good girl. Surely that's her own nose."

"Only kidding, mate. I think I'm switching from beer—have a couple of silver bullets instead."

"Not me. I can't handle martinis. They wipe me out."

"Have one with me. You've got nothing to lose, either." There was something very sour, almost bitter in Wick's inflection. Sherman decided to stick to beer. He waved off the martini the waiter proferred at Wick's insistence, and ordered beer instead. The music broke up to take a break.

Billy said that he'd seen Mariano walking through Knightsbridge near Harrod's. "He had a black eye—dirty great mouse under his left eye." They discussed the dangers inherent in mucking around in the demimonde of gambling.

"Big Chief Fuck-nose is going to have a nasty ending if he's not careful. Kaput. That's his problem. I'm going back to Australia next week to rest. Tired. I'm tired. Have to find a replacement for myself in the group. I'll rejoin them at Longwood. I like Boston. I've always played well at the Cricket Club. What about you?"

"Oh, I'm playing Quebec City and Bangor. Then I'm off a week. That will be a nice change."

"You know, speaking of Randy, you know his mother stabbed his father to death—then took a little nap in the gas oven? When he was eleven?"

"No . . . my God! No wonder!"

"Well, I don't know. We don't even know that it wasn't his fault. Look! There's little, pretty, Missy Teacup—just like a Barbie Doll tonight, our little love is."

"Why don't you two make it up tonight? You aren't still up-tight about her are you?"

"Not half mate! She's so flakey that the only way I'd have anything to do with her was if the prime minister ordered me on a mercy fuck."

"You're certainly all charm tonight, Wicko. I think I'll clear off before you corrupt me and spoil my evening altogether." Billy left and Wick sat down and diddled around on the piano, picking out a tune.

When Billy joined Laurie, Jesse, Shep, and Missy, they were having a hoot over strange signs they'd seen around the world. Shep had just told the Howard Hughes one. Jesse picked it up, "Up in Bretton Woods, I saw a tombstone in an old cemetery, you know, two, three hundred years old? On it it said: 'Sarah Godsby—Died a Virgin—15 years old—1722,' and someone had chalked under it: 'Who said you can't take it with you?' " Warm laughter.

Missy said, "I saw a cute one in a restaurant in Cape Cod. It went: 'Shoes are required to eat in the cafeteria,' and someone had added, 'But socks can eat anywhere they like.' "

Jaguar had joined them in the middle of the story. He had one for them; maybe two. "I was going across a bridge in New York named after the traitor, Washington, and a sign said: 'In Case of Atomic Attack, Drive Off Bridge!' Well, I'll be goddamned! I know what they meant but, you know, somebody might just do it! The other thing I want to ask you about is a sign I've seen in every post office

I've been in in the States. It says on the door: 'No dogs allowed except seeing-eye dogs.' Now—who the hell is *that* written there for?" There was general laughter and agreement on that one. Several gave feeble reasons for it, then laughed at their own reasons. They all turned toward the noise booming from the piano. Wick was pounding the keys in concert style, but with his fists. The piano player was trying to coax him off as the drummer and bass player took their positions.

"I think I'd better rescue that piano," Jag said. He and Jorge converged on it at the same time. Wick gave it up without a struggle and, plucking a drink randomly from the tray of a passing waiter, joined them. They watched his approach with commingled humor and hesitation. There were a few anxious faces among the group, which had grown larger. Wick walked up to Laurie and said, "Pick a number from one to five." She balked for an instant, then said, "Three."

"Wrong. Take your clothes off." Laughter. "You see—if you agree to play the game, you have to obey the rules."

Consuelo jumped on that immediately, "And don't forget, Al. That extends to the game of society, of social things too. Be good, please."

"Balls," Wick said, with a popping sound, and, asking "Where's the bog?" he lurched off to find another group—and drink.

Missy and Karen made plans to take Zoe, the Frasers' little girl, to the zoo the next day, Sunday. "What's Jesse doing, practicing?" Missy asked.

"Not all day, no. He and Phil, you know Phil Katz . . . ?"

"Not in the biblical sense," she chuckled.

"Well, they're going to look at real estate tomorrow afternoon."

"Are you settling here? Are they looking for a house?"

"Not quite. They're looking for a tax shelter."

"Right on. I'm all for that. When I'm old and my playing years are far behind me, I'd like to have a little bit of all the money I earned instead of giving it to the government to flush down the toilet."

"What do you think about Jag and Consuelo getting married? Kind of nuts, right?"

"I'm not so sure, Karen. He may be just 'bout ready. Consuelo? Well, I'm not so certain about that part of it. Maybe."

"Did you ever have any—eh—how shall I say? With Jaguar, I mean, did you . . . ?"

"Oh, sure. A couple of years ago we bounced around a little number. I know that shocks you, but our lives have been very different, yours and mine. Sex has never," she hesitated, remembering her doctor's advice, ". . . we just have different ideas about it, I would say, Karen. Mine are changing—changed now, though. Any animal can couple. There has to be love to make love. I hope Consuelo teaches him that."

"No good telling him. He'll have to learn for himself."

Jorge and Phil were at the far end of the room discussing players. "They have a disjointed way of talking, your tennis people. When there's a lull, they tell a joke. They aren't very good socially, are they?"

"Oh, now, George . . ."

"Hor-hay. My name is pronounced Hor-hay." They both laughed. Jorge offered Phil a good cigar, which he happily accepted.

"Like any other serious businessmen or entertainers," Phil continued, "they—what they really want to talk about is tennis, but they think that's gauche. But it's their business. Businessmen talk business, artists and politicians, all of them, talk about their work at parties, or about people—and they *all* tell jokes."

"There is an axiom that goes: 'Intelligent people discuss ideas, mediocre people discuss events, and boring or stupid people discuss people.' "

"Ah, ha," Phil continued, "We're talking about people—tennis people."

"No, *amigo*. We are discussing the idea, the essence of them. Quite different."

"They're good, intelligent people, Jorge."

"But I have no doubt about that, but you see, they bore me, *amigo*. And that, unfortunately, is that."

"You'll change your mind when you know them better."

"Perhaps, but permit me to doubt."

Angie joined the swelling group and Laurie drifted away. They despised each other at this point. There was a hot discussion of philosophy and life styles going on. Shep had just finished a short, tight dissertation on existentialism, summing it up by saying, ". . . and even though I believe everything in life is absurd and futile—life to me is still a miracle. I have to justify being the recipient of the incredible gift. Furthermore, I've always been torn between two shades of philosophy, and they are—like—a real, a tough dilemma. The first is an old Spanish proverb that goes: 'Whatever lives, be it a tree or a man or a bird, should be touched gently, for the time is short.' The other is: Life is so short that you shouldn't take any shit from anybody." Most of them laughed and nodded at hearing articulated something that they vaguely believed.

Missy said, "There should be more music in people's lives. Not because I'm a pianist, because I'm not anymore, but everything . . . when you believe that all arts strive to be music . . . everything is glorified, given more . . . oh . . . body, substance, by music. We need it in our lives, and more of it. It creates more love in life."

"Here, here, Missy," Billy Sherman, the roving philosopher put in. "Fuck hate. That is the irreducible minimum of positivism!"

"The *what?*" Karen asked.

"You don't expect me to be able to say that again, do you?"

Wick tried to turn the conversation into a theological trend of very dubious taste, asking, "Did Jesus fart or didn't he?" No one took him up on the God-Man argument. Karen, a deep-rooted Catholic, turned her back on him and seethed. He was getting very drunk. Putting his big arm around Billy, and thereby creating his own little captive audience, Wick said aloud, mainly for Missy's benefit, "Wickie, d'ya know why a cunt's called a cunt?"

"No, why?" came the straight line.

"Have you ever seen one?" Wick roared at his own joke.

Missy spun around, pronouncing crisply, "The language that crawls out of the primordial ooze of your mind!"

"My, my. Will you listen to the pukka memsahib here. Surely, with all the experience, all the men you've had, you heard language like that before?"

"Your stinking innuendoes—that double standard machismo crap —I've heard it all my life and I'm damn fed up with it! If a man had balled fifty girls, he's a great cocksman. If a girl had, she's a pig. Not that you'd know anything about it. You just *talk* a big game!" Missy was very angry and she was rapidly losing self-control.

"Listen, you slut—if you're implying . . ."

"That is quite enough!" Jesse interrupted, thumping his beer down on a table. "You're out of order, Wick. Don't use that kind of talk in front of my wife, or any other woman. That kind of stuff is rough even in the locker room."

"I quite agree. That won't do," Sherman said, releasing himself at the same time from Wick's clutches. "This is hardly the time or place for this sort of thing Wicko, you great, steaming nit."

"The same to you with brass monkeys on it. I was only kidding." Wick said this in a bantering mood, but his manner and tone changed in mid-sentence, and he became surly again. "But trying to be amusing around here is a bigger waste of time than kicking a zombie in the ass!" He left for the bar.

Laurie came back on Phil's arm. "What's all the noise over here?"

Jesse retrieved his beer. "Nothing, Laur. Wick lost his match—bad match, bad humor. He's gone on the piss. That's Wick. He'll get over it. He's a good guy, Wicko is. Let's go eat something, shall we? I'd like to have a reasonably early night—get home soon."

They straggled into the buffet. Jaguar wandered among them, asking generally, "Has anyone seen Consuelo?"

"She was with Angie. I heard them say they were going upstairs to look at some album," someone contributed.

"I'd better find her," Jag said with a smile. "Minor crisis." He headed off toward the staircase.

Missy and Karen exchanged glances. Karen asked, "Doesn't he know about Angie? I mean, isn't he worried?"

"His ego is ironclad. He trusts Consuelo. God knows, I don't know any reason why he shouldn't."

"Between you and me, Miss, why are there so many lesbians on the tour? I can't honestly think of any queers among all the guys."

"It's a tough, exacting game, Karen. Not just the sport, but the fight with the men for rights—for money—for recognition. The WTA, the women's union, has gradually changed all that, but still . . . I think, in competing with men, the girls, subtly at least, become like them—even to the point of taking other women away. I believe sincerely that some are lesbians out of pure spite. Also, you know, top athleticism demands an almost masculine stamina at the top, and girls with, what shall I say, an imbalance of hormones—more male ones—tend to be stronger and more dykey. The opposite holds true on the other side of the ledger. Men with too many female hormones or tendencies get into the provinces usual for the female, like interior decorating and hairdressing, that kind of thing."

"Everyone knows that, but look at other sports. Gymnastics and stuff. The Russian and German girls are strong as anything but very feminine. And what about you?"

Missy dropped her voice several octaves and answered, "What about me?" They giggled a bit. Missy continued, "Just take a look at women's golf."

"I don't know about that. You may be very wrong. Appearances are sometimes deceptive. What about skating? It's terribly physical but I never met one of those on my tour. At least I don't think so."

"Whatever. I hope Jag isn't in for trouble. Not that he doesn't deserve it."

"I understand just what you mean. He even made a pass at me, once. Jesse would kill him."

"He's a hell of a man, though. Whew!"

"Oh, really, Missy, stop that!"

The party was winding up. The guests were saying their good-byes—cars were screeching out of the huge driveway. The row of lights stretched down to the gates, moving slowly until they hit the main road. Jaguar and Consuelo appeared, deadpan, from the upper regions of the house and were receiving the thanks with tired, strained faces.

Shep was driving Karen and Laurie back in Jesse's car. Jesse himself was going with Wick to give him company, keep him awake, and eventually, with luck, get the wheel himself. Jorge had already gone to bed.

Laurie made herself fairly comfortable in the back of the small Triumph. She kept looking back for Wick and Jesse. They were to follow, caravan style, back to London, so Jesse could pick up his own car at the hotel. The car purred along in the velvet night; occasionally the red taillights of a car would break the night ahead, but otherwise it was lights on sleeping trees, hedges, and high banks. The only thing that passed them was a large van, squeaking past them on the narrow lane.

With great relief, Jesse realized that Wick was going to drive sensibly and he wouldn't have to wrestle the keys from him. Wick was quiet, whether because of embarrassment, fatigue, or too much drink, Jesse knew. Probably all those things. Because Wick couldn't find the right key, they had been late starting and were two cars behind Shep. Wick overtook the first one and passed it easily. It was a small MG.

Nevertheless, Jesse exhaled a large sigh of relief when they were back on the left-hand side. He couldn't make out the occupants, so intent was he on making it by. Wick wasn't weaving very noticeably, and laughed when he did, fully aware. That was a good sign.

Five minutes later they came upon a car that was unmistakably Freddie Moore's. Their beams picked out Freddie and Missy in the front seat. Wick hit the accelerator on the first, short, straight stretch of road he found and hauled out to pass, gunning full throttle. Expertly, he whipped around them, but didn't gear back or brake, wanting to put lots of space between them. Jesse saw the light first—the van curled around the tight curve, but Wick's reaction was slow. One great, ear-splitting noise of grating steel on steel, and pain. Wick's next recollection was of lying out in the road where he'd been thrown amid the smoke and debris. The van was on its side, the wheels still spinning. A few feet from him, Wick saw part of an arm. He recognized his own wrist watch on it. It was his last thought.

The hospital's depressing, antiseptic pungency, ether and disinfectant and pain, was particularly cloying to Jesse in his grief. He allowed his wheelchair to be pushed down the corridor, oblivious in his pain to the pretty white phantoms who floated by, their faces registering real concern for him.

Wick was still in a coma in the intensive care room. Tubes seemed to loop all over the room, from his nose, from bottles plugged into his remaining arm. Tears flooded Jesse's eyes at the incredible spectacle; the zigzagging lines across the oscillator, blinking red lights, and tubes, tubes everywhere. He simply could not bring himself to believe it—that was WICK lying there, Wicko, lying there somewhere between their old life together and death.

A Sister was leaving Wick's side on her way to another patient.

"You can go in if you like, but he's—sleeping," she said.

"Still ruddy unconscious, you mean," Jesse snapped, with uncharacteristic sharpness. He was ashamed. The words hung coldly in the air. "Will he be all right then? Is he going to make it?" Jesse said this with exaggerated softness, as if to erase the harshness of his first statement.

The Sister took both tones in stride. She was inured to the manic mood swings of the sick.

"With prayers and treatment, he has a very good chance. It won't do him any good or alter the outcome of anything if you get sick worrying about him. He's getting the best of care, and I wouldn't be the first to say that your Mr. Wick is as strong as an elephant." She touched his shoulder. "Don't worry."

For some seconds, Jesse stared through a tear-blurred frame at Wick, then nodded to his own nurse. She wheeled him around and back, through flashing lights and vague forms to his own room.

Jesse was released from the hospital Monday night. On Karen's arm, he was limping slightly and breathing with some difficulty. Five ribs on his right side were taped up. Two were cracked. With Karen's consent, he had decided to try his quarter-final match against Kerwin "I might as well have a go. I'm here . . . alive." The doctors only shook their heads. One of them compared it to a raging bull pawing the ties of a railroad track about to attack an oncoming deisel. "I admire his courage but I question his judgment."

Jesse's escape from the accident was a near miracle. Mostly, it was his safety belt and the side of the car he was on at impact. The ribs, a slightly bruised knee, and a bump on the head were incredibly minor. The shock that engulfed him was more serious. Al Wick's tragedy was still a half grasped, half rejected truth. The van driver was dead. They shared the blame equally among them—Wick, Jesse, and the other man. Wick and the van driver were both speeding. Jesse should never have agreed to permit Wick behind the wheel. He knew that Wick had seen the approaching lights, too, but hadn't hit the brakes until the very last moment, almost as if it were intentional.

Eduarde Vector, sitting behind the wheel of the car that was parked outside the hospital, was the only one who would ever know the truth about Wick's feeling for Jesse. It had been drunkenly en-

242

trusted to him one night in Sydney. Jesse was the only person Wick had ever loved. It never went beyond that, and was a barely conscious, uncrystallized thought, but nonetheless . . .

It is extremely difficult to play a person who is handicapped, that is, difficult for anyone with scruples and dignity. Kerwin had both, and consequently, with tentative, patchy play, he was down a set to Jesse—Jesse, who was toiling in great pain in spite of the aspirin, Darvons, and an injection. Sammy pulled his game together finally, when he realized that Jesse was getting to more balls than his injury should allow, but it was too late to save the second set.

Every forehand became a flash of white, dizzying pain, and a deep breath made Jesse's entire body shudder. A heart attack must feel like this, he thought. But all the hard work and invested emotion he had in this match, the goal of regaining his title—no, he just couldn't quit. Tactics were out the window now. Jesse was playing on raw guts and reflex actions. Slowly he became unaware of the crowd, the place, even in his desperate concentration on . . . just the next ball, and the next one . . .

Kerwin saw his golden opportunity and made the firm decision to grab it. Whenever possible, he moved Jesse over to his backhand side, then chipped short and wide to the agonizing forehand side. Sweat ran freely into Jesse's eyes, and at times he felt that surely his chest would explode. He bore no malice because of Sammy's strategy. That was the game.

The drama wasn't lost on the fans, who winced and ached along with Jesse on every shot. He was their total focal point. They heaved and groaned for and with him as a shot would shrivel his face in agony.

Karen couldn't bear it any longer. She left the players' seats, her face buried in a wet handkerchief. Eduarde Vector, of two minds for an instant, started to go after her, but then turned and whispered to an ashen-faced Aunt Doobey. With obvious relief, she stood up and made her way hurriedly past the knees of the other spectators, bumping along and mumbling apologies. Those in the tiers behind —not knowing, of course, who she was—hissed, "Sit down!" "Can't you wait?" "What the hell's going on down there?" Or they just sighed angrily at the interrupted drama.

Vector sat down again. His pain was twofold: the empathy with Jesse, and the humiliation that he secretly wanted Jesse to play on in spite of the terrible pain. It was selfish and Eduarde Vector knew it. He resumed a kind of vigil with flushed chagrin.

At 2-all in the third set, Jesse clutched his side for a moment, before cranking off an ace. As Kerwin changed service courts, shaking his head, Jesse walked slowly, solemnly, but not without dignity up to the net and offered his hand. Sammy nodded, trotted slowly up to the net, and accepted the match. The crowd cheered, but the cheer was disappointed, heartfelt tribute to a courageous effort. Jesse knew that even if he did win the third set, he could never put himself through that agony again for another match. As Newcombe used to say, "If you play when you aren't fit enough to win, you're cheating everyone, especially yourself." With the applause ringing in his ears, Jesse, with a grunt, waved to the center-court crowd for the last time. Vector started out of the stands to help him off, but Jesse waved him away with a crooked smile. Kerwin walked slowly, a step behind him as Jesse made his little bow to the royal box and the British Patron of tennis. He looked back for a moment, but only for a moment. He was in too much pain for nostalgia.

Jaguar, on his way out to do a hatchet job on de Leon, paused a second to gently shake Jesse's hand. Jaguar, who felt everything deeply, was very moved. His nod to Kerwin conveyed a combination of "Well done" and "You've had it now." He was meeting Sammy next, in the semifinals.

Over on court one, Shep was doing a heavy number on Freddie Moore, hitting with determination and power. He couldn't figure out what was happening on center court. Jesse was two up, that he knew, but they couldn't possibly have finished the third set, so what was all the applause? On the changeover in his own third set, at 4-1, Laurie sent a note by ball boy: she had beaten Belinda Martin, and Jesse had packed it in.

Shep took the field, trying, as he had been for the entire match, to make power tennis obliterate the encroaching memories of the smells, sounds, and sights of the horror, the carnage, the . . .

He shrugged it off and gave full concentration to returning Freddie's big serve. Often, just at a volley, or on the toss of the ball for a serve, an entire film clip of the accident would flash before his eyes. It was Shep who, hearing the crash a mile away behind him, had returned, done all of the first aid necessary, comforted Karen, attended to Laurie, who became very sick, sent **Freddie and Missy** for an ambulance and the police—everything. It was a shocking thing for Shep. Wick hanging by a thread—his career finished. Just like that. It reinforced his own belief in the absurd, of the futility of everything, and he was taking it out on Freddie now. The momentary lapses didn't bail Moore out at all. Shep wiped him off the court, 6-1, 6-4, 6-2.

The quarter-final match between Jean-Claude and Mariano was a marathon five-set study in misery and bad feeling. Randy's protestations were illogical and desperate. He queried call after call with white-faced, thin-lipped intensity. He wasn't even trying to psyche Jean-Claude; Randy was literally fighting for his life. Something of that desperation was imparted to Dumée, who, through a subconscious spot of compassion, relaxed a little, just that imperceptible bit, but enough. Mariano proffered a limp, shaking hand to Jean-Claude after serving the last point for a 9-7 win in the last set. There were no words; only nods. The applause was sporadic because, although everyone appreciated the struggle, they felt, almost one hundred percent, that the wrong man had won.

In the bar later, Randy realistically appraised the moment of his hang-tough win. "I'm just lucky they didn't throw Irish confetti"— by which he meant bricks—"at me. The frog finally folded like an accordion." When Jean-Claude was told Randy's remark, he rebutted with cool hauteur, *"Quel culot, alors! Quel salaud! Il n'est q'une espèce de con, vous savez.* He stole the match, and I was guilty of contributing to his delinquency. He . . . I believe . . . he has, how you say, conned to me?"

The press conference continued, "Where do you go next, Jean-Claude?"

"I go to Cologne for a WCT *tournois,* then to Paris for International Team Tennis. I had a good run today, no? *J'ai failli* . . . I very nearly did it, no?"

"You almost did it. Good luck, Jean."

Jaguar came out of his press interview and headed for the players' enclosure, signing autographs by the handfuls as he labored along. It took him eight minutes to go twenty yards. Jag was relieved to squeeze past the pensioniers, the guards at the foot of the stairs, and climb to relative safety. He pulled up short with a rush of revulsion when he saw Consuelo, Angie, and Jorge sitting around in front of the television set balancing their drinks on the arms of overstuffed chairs. Jaguar was fed up with the whole putrid ménage, but he loved Consuelo—of that he was sure. Jorge was her cousin, and maybe he actually was somehow a decent bloke, and maybe, just maybe, Angie was there in order to be with him. The years had thinned out Jag's farmboy, wide-eyed acceptance of appearances, yet he still was host to a larger share of credulity than most of the other guys, whom he considered cynics. He tried to wash down his filthy, fearful suspicions with Consuelo's Campari soda. Ugh! It tasted like medicine. He couldn't understand how she could drink it. She reached up and kissed him with feeling, and Jaguar, with a heavy shrug, shook off his ugly vibrations.

"Let's have a victory round. We all did some nice work today." Jag walked over to the bar. He accepted the compliments and the liquor with equal attention. For a moment, he stood there looking at himself in the mirror. Since when hadn't he trusted his intuitions? There they were, behind him, sitting around like a wake because they were meeting each other in the semifinals. Oh, they were friends and doubles partners, but they had reached the semis in that event already; that surely ought to be consolation enough for them. Angie wasn't in Consuelo's league. No. Jaguar's gut inclination was to cut his losses, leave love to the movies, and play tennis. Wick's accident had shaken him out of his warm web. He had made up his mind to change it—until Consuelo touched his cheek. Slowly, her nails glided along his neck and scratched his hair just above the nape. Was it love or the fact that she wouldn't sleep with him? Jag sat down in a puddle of confusion.

After signing dozens of autographs, Shep and Laurie remained on a bench outside the court. They had won a particularly easy mixed doubles and were relaxing, allowing the crowd to dissipate before trying to bustle back to the dressing rooms.

"Real show biz, isn't it, Laur?"

"Superstars, the two of us. Bullshit, right? Do you feel like a superstar?"

"Sometimes. I try not to, but it's a sweet temptation."

"Just ride with it. What's that old song? 'If everybody were somebody, nobody would be anybody.' " Shep nodded with a smile and released himself to calm, to silence.

"Very peaceful out here," Laurie observed. They looked off, up the hill in the distance where lovely English homes were sketched into a canvas of suburban greenery.

"Charlie and Mary Hardwick-Hare live in that lovely, big white house over there," Laurie pointed up somewhere. Peace or no peace, Shep was bedeviled by his own indecision about Laurie. He was relieved when a crowd of well-wishers and autograph hunters discovered them and the arms, the hands with pieces of paper, programs, and pens jutted out at them like the goddess Kali. A thicket of arms surrounded them, and they had to sign their way out.

Exhausted, they separated at the dressing rooms with plans to meet in the Gloucester bar. On his way out to the car, Shep stopped a moment to watch the outside electric scoreboard, which was just then producing a ladies' double score instants after the applause for a point inside. Belinda Martin and another English girl were looking like winners. "Hooray for England," Shep thought, "glad to see them doing well. Just as long as they didn't break their dry spell in the men's singles."

The phone went off while he was on the john and he let it ring. Later, the roar of the shower water drowned out the knocking at the door. When he finally turned the water off and the din continued, he realized that he had somehow been aware of it all along. The racket persisted.

"All right, goddamnit, all right! Coming!" Shep slipped into a blue terrycloth robe and went to the door, drying his hair with a large towel. He opened it a crack. "Yes, what is it? What's the big deal?" Then, with some surprise, he stood back and permitted Randy Mariano to pass into the small suite.

"Naturally, I'm honored. What the fuck do you want?"

"Give me a minute, Shep. This is one of the hardest things I've ever done in my life. I'm here to beg a favor of you." Randy's voice quivered. He was obviously badly shaken by something.

"Want a drink?"

"Really. Yeah—that would be fantastic. I need it."

"Whiskey?"

"Straight, please. Far out."

"Okay, what's the problem? I'm in a hurry."

"I won't take much time. You can give me a yes or no. It'll only take that long. Only a minute."

"Randy, when I'm with you the minutes fly like hours. What's the hype?"

Randy spoke slowly, deliberately, and it was all too apparent by his intense delivery that he wasn't exaggerating. "Shep, this is a matter of my life."

"Get to it. I'm all prepared for something—something plenty distasteful."

"I'm in debt so deep that I can't possibly pay off in money. Shep —this is the deal—no, not the deal, it's only a one-way street—the offer for my life. The odds are that you're gonna beat me—no—wait a minute! Let me finish." Shep had spun on his bare heel and was pounding toward the wardrobe. "Shep—I'm not . . . I'm not asking you to tank it. It has to go five sets, that's all. Five sets—you win, Shep, you're the winner but the bet is for five sets."

Shep looked at him with olympian indifference. "Randy, you are beyond contempt. Do you know what you're saying? This is Wimbledon you're talking about. I don't mean for a second that I'd consider it anywhere else, but *Wimbledon?*"

"Goddamnit, Shep, we aren't rapping about Mecca or Jerusalem. Wimbledon isn't St. Patrick's Cathedral! Hey, buddy, they are going to kill me! Can you dig it? *Kill* me! All I'm asking—asking, shit, *begging,* is that you let me win two sets. I give you my word that I wouldn't make any attempt at grabbing the fifth set. For God's sake, Shep! I know you dudes always put the bad mouth on me. But think! Have I ever really harmed any of you? Maybe I was a pain in the ass, sure, but hurt anyone? Never. I never did. I'm not even sure I can't beat you, but I'll never know this time. Two sets, Shep."

"Look, Randy. A: You may be setting me up. You sound sincere, but you're a con man. You're very persuasive, yes, but that's your bag. B: If I have any standard at all, it's my own honesty. If I did what you want it might destroy me."

"I swear, Shep, I swear to you—I'm not lying." He tried to take Shep's hand. Step stepped back a pace. "You have nothing to lose. You win the match. Only two sets, for the love of Christ, Shep!"

"Randy, I'm going to try not to even think about it, and as of this moment, if you want two sets, you have to win two sets or the match or whatever. Now, get out! Fuck off out of here and don't try to contact me here or anyplace—ever! Get out!"

Randy's look of pleading disappeared. He straightened up, squared his shoulders, and left the room without another word. There was a slight, sad, twisted smile on his face.

Laurie was describing ninety-degree arcs with her revolving stool, slowly and dreamily. The music was sorrow-softened, something about unrequited love. She thought for a few minutes about her match with Missy the next day. Ten times, she had played it already in her mind. Now she was numb with it all. She decided to put it out of her mind, but absolutely. Whenever it would try to seep in, Laurie concentrated on Shep, or a white dot, to the exclusion of all else.

Shep swarmed over the stool next to her and kissed her ear lobe lightly. It gave Laurie a sharp jolt—whether love, or Shep dragging his feet on the carpet, she didn't question. "Sammy Kerwin called me. He's upstairs with his elbow in a bucket of ice. Wanted to know if I had a forearm support. Have you got one by any chance?"

Laurie hesitated a beat before answering. The question was well outside any of the frames of reference in her thoughts at that instant. "Ah . . . I may have one in my bag. There was one in there, I know, but maybe it got chucked out at some point. I'll look when I go up. It's adjustable."

"It's a pity you and Missy have to play each other in the semis. I'd much rather have seen you play the finals together."

The new shift in subject made Laurie certain that Shep had something more important to get to. Shep was doing what he himself called an intellectual soft-shoe. When she didn't carry farther with the subject, King's Gambit denied, he finally got to the meat and potatoes: Randy's visit. After swearing her to secrecy, he told her the shocker of a tale.

"What are you going to do? That's a hell of a bind you're in."

"What kind of question is that? What do you mean, what am I going to do? I'm going to beat him as badly and quickly as I can."

"Hadn't you better give it more . . . consideration? I mean, if he's telling the truth, don't you have a part to play? God . . . maybe they will kill him."

"I doubt it, but . . . I shouldn't have told you about it. My mind was already made up. Probably you are right. . . . No, I don't trust him."

He abruptly broke off that segment of conversation and fell head-long into another topic. "I have to admit the LTA and All-England Committee are doing a tremendous job here. It's a great tournament. Smooth. But it's a little pathetic. The unions and other groups have really reduced them. It's sort of embarrassing. They spawned all these wonderful, rebellious children. It's a bit like the primitive man whose straw house caught on fire and burnt his pig. It was the first roast pork. Everyone enjoyed it. But the hitch was that every time they wanted roast pork, they burned the house down. The unions and power groups didn't have to burn the house down. Sad, really. There must have been a better way out."

"What in the world is that all about?" It struck Laurie funny, this long monologue about the LTA, and she broke up laughing.

"I'm sorry, Laur. I guess it was pretty funny. It all just came over me. Couldn't this be an ordinary Wimbledon with nothing to sweat or think about but tennis? Shit! It's really upsetting. One horror after another. Do you mind if we don't make love tonight?" Third shift in text.

"Yes, it really is all right. You don't want to leave your game in the bedroom, right? I can dig it. But who's going to tune up my Stradivarius, hum?"

"You're using my own argument against me. That's fair, I suppose, okay. Just a little tuning. Missy was so hysterical about Wick it was unreal. I think she loved him. Well—he was *sui generis*. There aren't any others like him."

"He was a funny, sad man."

"Yeah."

"Shep, I don't mean to keep harping on it, but you do have to think out the Mariano thing all the way—all of its ramifications. Why do we hate him so? Why is he so bad?"

"It must be, now that my Latin comes back to me, what we used to say, *Corruptio optima pessima*—the worst is a corruption of the best. Like dirty, filthy snow in the cities is so loathsome, because the real nature, the essence of snow is to be white and pure. Randy is such a genius at tennis that his being such a rat-fink swine makes it that much more awful. Who gives a shit if mediocre people are bad?"

"I see what you mean. To change the subject again, I won't mind beating Missy. She always has the piano to go back to."

"Women are so illogical. Who was talking about music? Come on, let's go up."

Laurie secretly rolled her eyes to the ceiling and just shrugged her shoulders. "Remind me about the thing for Sammy."

No array of supports or supporters could have saved Sammy from Jaguar, who started out aggressively from the second game, after dropping his own service. Seventy-four minutes later it was all over, 6-3, 6-4, 6-2. Kerwin played well, but only as well as Jaguar allowed him to. Shep watched the match all the way, unusual for him, and was thoroughly alarmed at the peak of play that Jag was achieving. "I hope to God he's not holding back," was his comment on a TV interview for ITN, London Weekend Television.

In losing to Laurie, even in straight sets, Missy gave everything she had, put up a tremendous battle, and was rapturous to have come as close as she had. As she shook hands graciously with Laurie to a very pleased audience, she recalled for a split second that one year ago to the day she was sitting in an asylum with no world to go to. Even in defeat she was exhilarated. It rubbed off on Laurie, who felt even better about her win because of Missy's wonderful reaction.

On the way to watch Consuelo's match, Jaguar was called to the press room for a telephone call. It was Jesse.

"Hello, Jesse, you filthy beast. Whatcha want, mate?" Jag attempted a light note, but, always a victim of his emotions, got an immediate lump in his throat.

"Jaguar, you great git. Karen and I just wanted to say good-bye . . . and to ask you to give our good-byes to the boys—and the girls, of course. Karen nudged me. My Aunt Doobey's quite poorly, so we're—Vector, Karen, and meself—we're flying back to Sydney this afternoon."

"I saw a note like that once in an American Express book in Naples. 'Leaving Florence, going back to Sydney.' Got to be a pufta, right?"

"Very funny, very funny. Listen, Jago, good luck with the Yank, whichever one it is. Probably Shep. He's a good lad, but let's try and keep it in the Commonwealth."

"I'll have a go, Jess. Jess—I'm very sorry about Wick."

"Yeah." Silence.

"Well, all the best to Karen. Safe trip, Digger."

"Yeah, thanks—I think I'm retiring, but don't tell the press types."

"Balls. I've heard that song before."

"Well—we'll see. It's ninety-nine per cent sure, though, if I want to keep my happy home. And I don't want to burn out like Berconi."

"You'll do whatever's best, I'm sure of that. Bye-bye, Jesse."

"Bye."

Jaguar put the phone down slowly, then stared at it for a full minute.

"Finished with that?" he heard a voice saying.

"Oh, yes, sorry." Jaguar went up to the players' section, his mind sad and confused, overwhelmed by life's imponderables.

The first set went the way any bookmaker would have laid odds. Consuelo had too much class and power for Angie. The seeding committee had done a first-rate job. They couldn't have predicted that a qualifier, Kerwin, would have made the semifinals, or that Angie would have hit her best form after such a bad year. Nonetheless, her best form wasn't within reach of Consuelo's artistry. The second set was a travesty. It looked like a heavyweight holding up a middle-weight in the clinches. It was the waltz of the toreadors. Consuelo didn't put her away until it became almost embarrassing. After the final point of a 7-5 set that should have been 6-2, Consuelo ran up to the net with tears in her eyes. She gave two short sobs and they walked off with their arms around one another's shoulders; natural-enough-looking behavior to most, but to Jaguar, it was bright and clear. He was perceptive enough to understand that, not only was the semifinal over, his engagement was over as well. The part of him that had grown up was deeply hurt, but the blithe spirit that still abided in him was immeasurably relieved. Jaguar scratched his head, got reluctantly to his feet, and said, to no one in particular, "Fuck it."

Randy Mariano looked at Shep only once in the dressing room. Shep didn't see him, or avoided seeing him, and Randy couldn't read anything in his face. Randy's mouth was very dry. He put a mint in his mouth. It was easily a full minute before any saliva set it free from the roof of his mouth. For some reason he was chilly, and put a sleeveless sweater on before going out to court one.

Shep sprayed some resin on his grips and walked off after him. His thoughts tumbled like a short-circuited computer that never comes up with an answer.

If Mariano was nervous, he didn't exhibit it in the knock-up. He was hitting out at the ball and hitting deep right from the outset. Shep kept reminding himself that the match wasn't a "gimme," that Randy was most capable of a singular rip-off. And the first set proved him right. Mariano broke him in the seventh game to lead 5-3. Shep broke right back, only to lose the set in a tie-break.

Shep then pulled all the stops and played up to his stature, his capabilities, and hatred. He swept the next two sets with a dizzying array of power shots—heavy, blinding power. For two games, he let up in the fourth set. Randy led 2-1 with a break. It was an exercise in discipline for Shep, zipping up his mind, his imagination, against Randy's alleged predicament. At 4-3, with Randy serving, Shep hit a blistering backhand return of serve up the line. Randy tried to change up. Shep ran around it and hit another clean winner. A double fault and bad volley and Shep broke back. He held serve, then jumped on Randy's second services and broke him again. Mariano had blown. At 5-4, Shep was poised, ready to serve out. Randy wasn't set yet. Shep stepped away, like a batter when the pitcher takes too much time. Mariano's face had gone white. As he stepped into position, he dropped his racquet. He seemed to have atrophied. Shep fought for concentration and hit a serve for an ace. It wasn't that good a serve. Mariano didn't seem there. Shep thought about asking him if he were ready, but knew the answer. He restrained himself from the beau geste. After two long rallies, Shep was at match point. Randy's face was pleading. Shep went into his wind-up, but Randy signaled and called that he wasn't ready. Shep bounced the ball, went into his motion, and boomed an angled serve short to the backhand. Randy pulled off one of his famous chip shots, but Shep, with a burst of speed, chip-volleyed it back at an ungettable angle. It was all over. The brilliant shot brought the crowd to its feet. They were also raising a clamor because the good guy won and the villain got his just desserts, tra-la-la.

Randy walked up to the net. His hand was trembling, but his grip was firm. He looked Shep straight in the eye, and with a weak grin, said, "Well played." His eyes were very moist. That evening on television, Shep commented on the fine umpiring in the tournament, and mentioned, in passing, that Mariano had conducted himself like a gentleman. When they had asked Randy to appear on the same program, he responded, "Why don't you smoke my pole!"

The morning had been fresh and bright for the ladies' final, but at about one o'clock it became Black Friday as dark, swooping clouds blotted out the sun. A sharp, gusty wind accompanied the clouds, adding a somber sound to the proceedings as it whistled through the royal box canopy. For many, Laurie was the underdog, because Consuelo had won twice there in previous years. Laurie was a crowd favorite because she was young and seeking her place in the sun, wanting to have her chance to be indelibly, immutably inscribed in the book of Wimbledon's winners—*the* golden book of tennis. Consuelo had her following, too. Not only her countrymen and other Latins, but the Old Guard who wanted her to win for the old folks. Consuelo was twenty-eight.

The first four games went against service as they both started nervously. The agony of suspense was maintained at a breakdown pitch, as they seesawed in the lead. Consuelo won the first set—the only set Laurie had thus far dropped in the tournament. Just as tightly, Laurie eked out the second set, breaking serve on the last game, the only service lost in the set, for 7-5. Midway in the third set, it was obvious to all that Consuelo was tiring and limping from time to time. Laurie noticed it and went into a campaign of running her from side to side regardless of who won the point, and dropshotting her even from the back court; anything to make her run. With Consuelo serving from 3-4 down, it was over. Laurie had taken it all out of her and the serves were half-paced and short. As Pancho Segura advised, she ran around her backhand and went for winners off the forehand. The serve was broken, and, without losing momentum, Laurie concentrated like a horse with blinkers pulling a load and served flat out, humping big boomers into a very tired Consuelo, who was almost helpless. It was a brilliant, brutal display of raw power. Laurie Silverman was Wimbledon Ladies Champion. Her tears were light and quick-dry. The carpets were laid and the ball boys lined up on either side of them as the Duchess of Kent made the climb down from the royal box. Photographers were everywhere. They sounded like a million crickets on a quiet night. Laurie looked around at the applauding mass of faces as if in a dream. She couldn't recall one word that was spoken between the Duchess and herself—only a kind, pretty face beaming goodwill on her. While Consuelo was receiving the runner-up trophy, Laurie slowly drifted back to earth. She muttered to herself, almost audibly, "Jesus Christ, I did it!" As they made their bows and walked out of the arena with the photographers following, clicking away, there could have been celestial music playing for all she knew.

The Silvermans laid on a sumptuous victory dinner for Laurie and her friends—and quite a few of their own—at Aretusa's. Shep permitted himself only one glass of champagne, very little of the heavy, but delicious, Italian food (he was too nervous), and passed up the nightcap dancing at Annabel's completely. Saturday was his big, all-important day, and it would be dawning for him in just seven hours. Laurie had a ladies doubles final the next afternoon, but thought that she should put in a short appearance for the family and group at Annabel's. Shep drove her over to Berkeley Square, and they waited in the car for the doorman to come.

"I have to tell you once more, one more time, how magnificent you were, my Annie Laurie."

"Thank you, babycakes. I love you." For an answer, he kissed her tenderly. She waited for him to say something, and when it was certain that nothing was forthcoming, she released a sad sign of resignation. "You don't, do you?"

"I just really, honestly, don't know, Laur. It's all too quick. I'm older and my hide is tougher, I'm sorry to say. Maybe, just maybe I do. I don't mean that to sound as if I'm doing you a favor. Can we talk about it tomorrow night?"

"Oh, sure. I'm sorry. I got caught in the riptide of my own sweet . . . oh, it is very sweet, Shep! I did it!"

"You certainly did that. Here's Joe. I'll walk you inside. Don't stay long. Be out in just a sec, Joe!"

Shep turned her over to the headwaiter, who led her off to her table. Then he turned to his favorite bartender and said, "Keep an eye on her, Sydney, and in half an hour take her a note that I said to wrap it up and get to bed." He gave Sydney a pound note and left for the quiet—and loneliness—of his own dark cave at the hotel.

Once upstairs, in robe and slippers, he took a can of beer out of the tiny fridge, and sat down in the dark. For a few minutes, he went over his game plan—making adjustments for the new virtuosity that

Jaguar was showing. On form—that is, form up to that moment—it was his, Shep's, Wimbledon. But Jag was at peak performance and not to be underestimated. He was playing like a bitch. Jaguar had seemed a different person in the last couple of days, and he'd diffidently introduced his parents that afternoon. "These are my parents, Mr. and Mrs. Gray," had been said too shyly for Shep to laugh. He laughed now. Jaguar was a strange one. Strange and simple at the same moment. He had confided in Shep—Shep, his antagonist at the finals—that he was breaking off his engagement. "Bloody shame, really. She had dirty, great nipples! Lovely, they were." When Shep asked him if Angie figured at all in his decision, he nodded. "I'm going to give Angie a poke tonight just to be rude. Wouldn't that be famous?"

"Ironic. Just beautiful," Shep had agreed.

Being in different groups, they hadn't played each other often through the years and, at that, not for a long time. Shep had scouted him well and didn't expect any surprises. If only he himself played as well as he could tomorrow, of all days!

His first essay at sleep was a failure. All of the possibilities were little vignettes linking the shadowy hours into a chain of anxiety. Sleep was abrupt and heavy when it ultimately came.

Although the air was festive enough when Shep pulled up at the dressing room door, the outside courts, empty, quiet, seemed like a cemetery. All of the hopes, hard work, and wounds, the aspirations and fears of the faceless army of challengers were buried out there.

There was a slight wind and the pennants snapped from time to time in short gust-spurts. Things were more orderly on finals day, the crowd smaller. Only those who could fit into every seat in center court were there. The weather report said in true British fashion, "Occasionally clear patches, variable, with a possibility of rain." How about that for fancy dancing?

Although Shep had showered at Queen's after his half-hour warm-up, he prepared to take another, his theory being that the hot water loosened up the muscles and warmed one up faster. Billy Sherman, the little bluebird of happiness, ran up to him, his face like the BBC with news of fresh disasters.

"Jesus, Shep! Did you hear what happened to Mariano?"

Shep threw a sock in his face and stepped into the shower. He blanked Randy out of his thoughts. Jaguar came in from an outside court to take his shower. He was eating a honey-and-peanut-butter sandwich and drinking light cider. No clues or tips for aspiring young players there.

At two o'clock sharp, they patted each other on the back, not overly fondly, and followed Dad's Army out to the vast center court. Shep and Jaguar weren't close friends but they had come a long way up together from those fledgling days in the Caribbean. Their friendship was like parallel tracks that go off to the horizon and seem to touch but never actually do.

They made their bows and got themselves together at the umpire's chair, both of them marshaling all the energies and input of their tennis careers. "Linesmen ready? Players ready? PLAY."

The first set was tentative to a fault. Both were hesitating on their shots—not going for their chances. After exchanging service breaks, Shep broke the faltering service of Jaguar and went on to win the first set 6-4. It had been a long set, and at that rate it was going to be a long day.

Jaguar started to hum into action and pulled his game into a more efficient machine. His volleys were penetrating deeper and his ground strokes were picking up authority. Shep had a bad service game at 4-5, shouting "Concentrate!" at himself—but lose it he did to give Jag the game and eventually, after some patchy serving by Jaguar too, the second set.

The third set was better. Better for both of them. It picked up in intensity and had more rallies with sharp, expert drives by both of them. They were suddenly showing why they were professionals, experts, the cream of the game. The crowd increased its emotional pitch right along with them. The partisan weight of the British was hopelessly top-heavy to Jaguar, for whom they roared their ecstasy on his every good shot. Shep got a nice ripple of polite applause on his good shots, and a few Americans in isolated sections were cheering their guts out like thrushes in a tornado.

Jaguar was behind on his service, 15-40 at 5-6. He ripped a blazing flat serve past the flat-footed Shep for a flat ace. Shep never saw it; at least he couldn't remember seeing it when he tried to recollect it later. The next serve was another cruncher to the backhand that Shep whipped wildly—and luckily—cross-court for an outright winner and the third set.

Jaguar's face was impassive as he braced to receive serve in the fourth set. The game went to four deuces before Shep, with great relief, cracked a backhand volley cross-court to win it. There was less gusto from the audience now. They were immobile banks of color in the stands. Short blasts of wind made the canvas over the royal box snap angrily.

Midway through the fourth set, Jaguar seemed to shift gears and play more smoothly. He was getting a jolt of confidence from someplace. But Jaguar's improved play was raising Shep's. It was as though two highly professional actors were delivering lines back and forth, bringing out the best of each other. One super shot brought a flash of brilliance from the other. It was flawless, errorless, exciting play rising in a felt crescendo.

They changed over at 3-4, going with serve. Jaguar threw down his racquet and unzipped the cover on another. He tapped it on the palm of his hand twice and picked at two center strings like a banjo.

Jag was now serving with rhythm, everything meshing, but Shep's concentration was beamed—absolute. "Watch the ball, move your feet," he muttered almost out loud. He felt good—aware of his every cell, muscle—super-ready.

It was powerful stuff in the real sense of power. A direct line of some sort of electricity bound them together in high-voltage, dramatic, head-to-head struggle. The ripples of it drifted through the center court crowd and they themselves became plugged into the current.

Shep wanted to break this one for a 5-3 lead to serve for the match. Jaguar ran off to 30-love, but two good returns and a loose volley by Jag gave Shep break-point. Shep hunkered down close to the ground, trying to sight the incoming ball on its own line of trajectory, like a pool ball. Fragments of his acceptance speech flickered through his mind but he banished them . . . brushed them aside quickly. He wasn't going to make that mistake.

Jaguar hammered a serve so hard to Shep's backhand that, even diving and blocking, he never touched it. Another unreturnable serve and a groundstroke battle that ended in a touch-stop volley gave the game to Jaguar. For the time being, the momentum was his. Jag stayed on top of his mental advantage, broke service and decisively served out, and it was—two sets all.

At the start of the fifth set, the sun went into the clouds and the warmth was slowly sucked up until the air became brittle with chill.

Although the crowd became alive with pulling on coats and sweaters, the players barely registered the change and battled on, dead even in shot-making and points. It was gradually more apparent that Jaguar was using more subtlety, more touch shots as the match went on. He would suddenly break up the little string concerto and resume thumping the ball. The fans were electrified with the boggling display of great shots met by better, back and forth, and after every Jaguar point, the clamor was deafening. There was the clear, wondrous possibility of a British winner after a long drought.

Shep was somewhat nettled by the partisanship, even though he was a psychologist, understood it, and was fully aware that there was no malevolence involved—well, very little.

Jaguar, serving at 5-5, was down 30-40, and the crowd was tombstone silent. A net cord return of Shep's spun on the net interminably. Jaguar's face was frozen in a study of horror and helplessness. The ball fell back on Shep's side, and over twenty thousand people began to breathe again. Given this new life, Jaguar took a deep breath himself and served out for 5-6. Having been so close to the service break made it seem like a reality denied, and upset his sense of proportion. Fatigue crept into the machinery, too. On his own serve, Shep overhit a smash, missed a short volley, and, after walloping an unreturnable serve, sent a routine chip approach shot long. 15-40. Shep inhaled deeply. He tried not to dwell on the doomsday aspects of the next point.

He bounced the ball five times, reared back, and hit a kicker. It bounced high, but Jag jumped around and whipped a topspin forehand well out of the reach of the diving volleyer. Shep was still down like a fighter enduring a surprise knockout and listening to the count. Game, set, and match to Gray, 4-6, 6-4, 5-7, 7-5, 7-5.

He looked up from his knees as Jaguar sent his racquet six stories into the air with an Indian war-whoop, and rushed up to the net to shake hands. It was pandemonium. The English were delirious and Jaguar was in a mad rapture. Shep, with a polite, shattered smile, shook his hand and, brushing the grass from his shirt, walked abjectly to the sidelines like a man who had just been condemned to death. During the presentation ceremonies, they were both in a daze —different dazes; Jag's, warm and milky; Shep's, ice cold. He waved weakly and smiled to Laurie who was weeping freely. His mother's brave smile was even more heartrending to him. With a faint voice he whispered, "Sorry, Dad," and almost cried. As he watched the cameraman fawning and snapping around Jaguar, who was holding the trophy in the air, agleam with smiles of total joy, Shep was numb from his toes to the roots of his hair with disbelief. It was Jag who came and put his arm around his waist when it was time for them to march out. Jaguar's happiness turned to heavy disappointment for Shep even as they walked off. Strange guy, this Jaguar, Shep thought. "Cheer up, Jag, you won. Don't pull such a long face. You're the champion, buddy." They both laughed, and the crowd gave them both a three-cheers, hip-hip-hooray, for good sportsmanship.

Well, what the hell, Shep thought, after that stirring tribute, I tried my best . . . oh, shit!

Laurie and Shep stopped by Jag's victory party for a brief, token appearance, had a quiet dinner, and parted. Laurie had to take her parents to the airport that night and Shep didn't feel like company anyway. They decided to meet the next day in Hyde Park before Laurie had to catch her own plane. "Where do you go next, Shep?"

"I'm playing at Cologne. You can reach me at the Esso Motor Hotel there. But I'll . . . I can tell you all that tomorrow."

"Oh, God! Did anyone tell you about Mariano?"

"No, and I don't want to hear it just now, Laur. I'll see you tomorrow. I have to show strength of character now for a few minutes at the television studio, then sleep, I hope." They kissed, less passionately than pitifully, good night.

Shyly, they met late the next afternoon at the Serpentine in Hyde Park. They drank a cold beer in the bar, then walked hand in hand, talking in the fearful tones of those who know the end of a relationship is near. They sat by the filmy, brackish lake. Shep tossed a stone among the lily pads, and the bullfrogs simmered down to a groundswell of gurgles for a moment, before resuming their throaty gusto. Laurie loosened her pony tail and shook her hair free, and there were glints of light in it from the lusty, wine-amber afterglow. Her face was taut with anticipation. The time for parting was near. Oh, they would see each other again, somewhere, but as different people. Somewhere a bird was prematurely making its absurd night sounds, a silly but lonely sound. Laurie, lost in thought, half consciously watched a slowly wafting hawk. Suddenly, it swept low over the tree tops toward them and startled her into sitting up abruptly.

At any other time, it would have been pleasant sitting there amid the pine and musk scents, sharing arcane secrets with the darkening trees, but now it was tense and dispiriting. Shep glanced at his watch. She saw and understood the gesture and they stood up.

In the growing dusk, they both looked down at the hands they were holding. Her eyes were moist and she couldn't speak. Shep was very moved, too. She looked up at him, and she was a very, very little girl. Hot tears seared her cheeks and she burst into choking sobs. She stopped as suddenly as she had started. "Never mind about me or Wimbledon, Shep. Next year's wine is the sweetest."

Laurie kissed him quickly and ran off in the direction of the road. He tried to call good-bye, but the lump in his throat prevented the anticlimax. The imprints of her tears were still damp on his cheeks as he looked after her and thought: Maybe that's right, Laurie. Next year's wine is the sweetest.

He stood up slowly, and, with an absent-minded, conditioned reflex, went through the motion of a forehand with the flat of his hand, knees slightly bent.